B. Earl
BSc, P.G.C.E.,C. Chem., M.R.S.C.

L.D.R Wilford
B Tech, PhD, P.G.C.E., C. Chem., F.R.S.C.

Thomas Nelson and Sons Ltd
Nelson House Mayfield Road
Walton-on-Thames Surrey
KT12 5PL UK

51 York Place
Edinburgh
EH1 3JD UK

Nelson Blackie
Westercleddens Road
Bishopbriggs
Glasgow
G64 2NZ UK

Thomas Nelson (Hong Kong) Ltd
Toppan Building 10/F
22a Westlands Road
Quarry Bay Hong Kong

Thomas Nelson Australia
102 Dodds Street
South Melbourne
Victoria 3205 Australia

Nelson Canada
1120 Birchmount Road
Scarborough Ontario
M1K 5G4 Canada

First published by Blackie and Son Ltd 1991
ISBN 0-216-93131-2

This edition published by Thomas Nelson and Sons Ltd 1992

ISBN 0-17-438632-X
NPN 9 8 7 6 5 4 3

Printed in Hong Kong.

CONTENTS

The 'Chemistry Data Book' contains the relevant data for courses leading to examinations in chemistry at all levels. The data is presented in a manner which makes it accessible to a wide range of students with differing needs. The data has been compiled from a number of standard international publications whose data is reliable.

Abbreviations, subscripts and superscripts

The following abbreviations, subscripts and superscripts have been used throughout the text.

Abbreviations	Subscripts	Superscripts
dec = decomposes	aq = aqueous	1 = nuclidic mass of an important isotope
ins = insoluble	g = gas	2 = uncertain value
liq = liquid	l = liquid	3 = under pressure
n.a. = not available	s = solid	4 = calculated value
sol = soluble		atm = atmospheres
T.M. = transition metal		BH = calculated via the Born–Haber cycle
ppm = parts per million		deh = dehydrates
m.pt. = melting point		eqm = equilibrium
		exp = explodes
		sub = sublimes
		+ − = positive and negative charges on ions
		$\ominus$ = standard

ELEMENTS

OF THE PERIODIC TABLE

Glossary of terms and units

This section contains individual information about each of the 109 presently recognised elements. The information given for each element is sub–divided into the following categories.

a) Its name and its derivation.

b) Atomic number and relative atomic mass (of the most stable isotope).

c) Its chemical symbol as well as the group and period to which it belongs.

d) The main source of the element and the history of its discovery. This section also includes details of a method of extraction of the element.

e) A brief description of the physical and chemical properties of the element.

f) Uses of the element and its compounds.

g) The biological role, if any, played by the element and its compounds.

h) The melting and the boiling points of the element (in Kelvin) along with its density measured in grams per cubic centimetre (g cm^{-3}) at 298K and 1 atmosphere pressure except where stated.

i) The element's electronic configuration as well as its known oxidation states. The principal oxidation states are highlighted.

j) The abundance of two stable or relatively stable isotopes.

k) The standard electrode potential, $E^{\ominus}$, measured in volts. This is the electrode potential for the element in contact with a 1M solution of its ions when measured relative to the standard hydrogen electrode at 298K and 1 atmosphere pressure.

l) The Pauling electronegativity value for each element. The Pauling scale measures the ability of an atom in a covalent bond to attract electrons to itself.

m) The radii measured in nanometres (nm):
   i) the ionic radius – this is the effective radius of the stated ion in a crystal lattice.
   ii) the atomic radius – this is half the distance between the centres of two adjacent atoms in the structure of the element.
   iii) the van der Waals radius – this is half the distance between the centres of two atoms in adjacent molecules.
   iv) the covalent radius – this is half the distance between the centres of two adjacent atoms in a close–packed structure or in a molecule.

n) Molar enthalpy data measured in kilojoules per mole ($kJ\ mol^{-1}$) except in the case of standard molar entropy data where the units are Joules per Kelvin per mole ($J\ K^{-1} mol^{-1}$):

i) 1st ionisation energy – this is defined as the energy required for the process

$$M_{(g)} \longrightarrow M^{+}_{(g)} + e^{-}$$

ii) 2nd ionisation energy – this is defined as the energy required for the process

$$M^{+}_{(g)} \longrightarrow M^{2+}_{(g)} + e^{-}$$

iii) 1st electron affinity – each value included represents the energy change for the process

$$M_{(g)} + e^{-} \longrightarrow M^{-}_{(g)}$$

iv) 2nd electron affinity – each value included represents the energy change for the process

$$M^{-}_{(g)} + e^{-} \longrightarrow M^{2-}_{(g)}$$

(Values of electron affinities do vary considerably according to the source used.)

v) Standard molar enthalpy of fusion ($\Delta H^{\ominus}_{Fusion}$) – this is defined as the energy required for the process

$$M_{(s)} \longrightarrow M_{(l)}$$

at a pressure of 1 atmosphere.

vi) Standard molar enthalpy of vaporisation ($\Delta H^{\ominus}_{Vaporisation}$) – this is defined as the energy required for the process

$$M_{(l)} \longrightarrow M_{(g)}$$

at a pressure of 1 atmosphere.

vii) Standard molar enthalpy of atomisation ($\Delta H^{\ominus}_{Atomisation}$) – this is defined as the energy required to form one mole of gaseous atoms from the element in the defined physical state at 298K and 1 atmosphere pressure.

viii) Standard molar entropy ($S^{\ominus}$) – this value gives a measure of the degree of disorder within any system. It is defined as the entropy of 1 mole of a substance in its standard state at 298K and 1 atmosphere pressure. In general $S^{\ominus}_{(g)} > S^{\ominus}_{(l)} > S^{\ominus}_{(s)}$.

o) The associated bond lengths measured in nanometres (nm), and associated bond energies measured in kilojoules per mol ($kJ\ mol^{-1}$).

Note

The standard molar Gibbs free energy of formation ($\Delta G^{\ominus}_{f}$) of elements in their standard states is zero.

Actinium

[Greek, 'aktinos' = ray]

Atomic Number	89
Relative Atomic Mass	227.0278[1]

Chemical Symbol	Ac	Group	–	Period	7 Actinide series

Main Source & History

Discovered by A. Debierne in 1899, Paris, France. It is naturally present, in small quantities, in ores of uranium. It is separated by ion exchange or solvent extraction. It can be artificially made by neutron bombardment of ^{226}Ra.

Properties

It is a soft, silvery white, radioactive metal. It glows in the dark. The most stable isotope, ^{227}Ac has a $T_{1/2}$ of 21.6 years. It reacts with water, hydrogen gas being evolved.

Uses	Actinium has no known uses.

Biological Role	Actinium has no known biological role. It is hazardous to health due to its radioactive nature.

Melting Point/K	1323[2]	Boiling Point/K	4373[2]	Density/g cm^{-3}	10.06

Electronic Configuration	[Rn] $6d^1$, $7s^2$

Oxidation State(s)	0,+3	Isotopes	^{227}Ac(trace), ^{228}Ac(trace)

S.E.P.	$E^\ominus$/Volts
Ac^{3+}|Ac	−2.13 Acid solution

Electronegativity (Pauling)	1.1

Radii	nm
Ionic	0.118 (+3)
Atomic	0.200[2]
Van der Waals	–
Covalent	–

Enthalpy Data	kJ mol^{-1}
1st Ionisation Energy	669
2nd Ionisation Energy	1170
1st Electron Affinity	–
2nd Electron Affinity	–
Enthalpy of Fusion, $\Delta H^\ominus_{Fusion}$	14.2
Enthalpy of Vaporisation, $\Delta H^\ominus_{Vaporisation}$	293.0
Enthalpy of Atomisation, $\Delta H^\ominus_{Atomisation}$	405.9
	J $K^{-1}mol^{-1}$
Standard Entropy, $S^\ominus$	56.5

Associated Bond Lengths (nm) & Energies (kJ mol^{-1})	n.a.

Aluminium

[Latin, 'alumen' = alum]

Atomic Number	13
Relative Atomic Mass	26.9815

Chemical Symbol	Al	Group	3	Period	3

Main Source & History

Discovered by H.C. Oersted in 1825, Copenhagen, Denmark, but first isolated by F. Wöhler in 1827. The economic production was made possible by C.M. Hall in 1886. Aluminium occurs naturally as bauxite (Al_2O_3). To obtain aluminium, pure Al_2O_3 is electrolysed as a solution in molten cryolite.

Properties

It is a hard, strong, bluish–white metal of low density. It is ductile and malleable and a good conductor of heat and electricity. Aluminium is a reactive metal, however, its oxide coating prevents it from reacting with air and water. It will react with concentrated, non–oxidising acids and concentrated alkali.

Uses

The metal and its alloys are extremely versatile and are used in applications requiring light weight and strength, for example, in aircraft, overhead power cables, engineering, kitchen utensils and cars. Thin sheets of aluminium are used as a wrapping material and to keep prematurely born babies warm by reflecting back lost body heat. Compounds of aluminium are used as catalysts ($AlCl_3$, Friedel–Crafts), in water purification ($Al_2(SO_4)_3$) and in ceramics and refractories (Al_2O_3).

Biological Role

No known biological role but aluminium accumulates in the body (approximately 20 mg per day) and is implicated as the main cause of Alzheimer's Disease (senile dementia).

Melting Point/K	933.5	Boiling Point/K	2740.0	Density/g cm^{-3}	2.70

Electronic Configuration	[Ne] $3s^2 3p^1$

Oxidation State(s)	0, +1, $\underline{+3}$	Isotopes	^{27}Al(100%)

S.E.P.	$E^{\ominus}$/Volts
$Al^{3+}|Al$	– 1.66

Electronegativity (Pauling)	1.5

Radii	nm
Ionic	0.054(+3)
Atomic	0.143
Van der Waals	0.205
Covalent	0.125

Enthalpy Data	kJ mol^{-1}
1st Ionisation Energy	577
2nd Ionisation Energy	1817
1st Electron Affinity	–26.0
2nd Electron Affinity	–
Enthalpy of Fusion, $\Delta H^{\ominus}_{Fusion}$	10.7
Enthalpy of Vaporisation, $\Delta H^{\ominus}_{Vaporisation}$	293.7
Enthalpy of Atomisation, $\Delta H^{\ominus}_{Atomisation}$	326.4
	J $K^{-1}mol^{-1}$
Standard Entropy, $S^{\ominus}$	28.3

Associated Bond Lengths (nm) & Energies (kJ mol^{-1})	Al–Cl(0.206,498), Al–F(0.163,665), Al–H(0.170,285), Al–O(0.162,585)

Americium

[After America]

Atomic Number	95
Relative Atomic Mass	243.0614[1]

Chemical Symbol	Am	Group	–	Period	7 Actinide series

Main Source & History	Properties
Discovered by A. Ghiorso, R.A. James, L.O. Morgan and G.T. Seaborg in 1944, Chicago, USA. The most stable isotopes are formed by multiple neutron irradiation of ^{239}Pu, and are purified by ion exchange. Metallic americium is obtained by reducing AmF_3 with barium.	It is a silvery, radioactive metal. The most stable isotope ^{243}Am has a $T_{1/2}$ of approximately 7400 years. It is an electropositive metal which reacts with air, steam and acids.

Uses	^{243}Am is used as a neutron target to produce higher actinides.

Biological Role	Americium has no known biological role. It is hazardous to health due to its radioactive nature.

Melting Point/K	1267.0	Boiling Point/K	2880.0	Density/g cm^{-3}	13.67^{293K}

Electronic Configuration	$[Rn]\ 5f^7 7s^2$

Oxidation State(s)	0, +2, $\underline{\pm 3}$, +4, +5, +6	Isotopes	^{241}Am(trace), ^{243}Am(trace)

S.E.P.	$E^\ominus$/Volts
Am^{3+}|Am	−2.07 (Acid solution)

Electronegativity (Pauling)	1.3

Radii	nm
Ionic	0.107 (+3)
Atomic	0.184
Van der Waals	–
Covalent	–

Enthalpy Data	kJ mol^{-1}
1st Ionisation Energy	578
2nd Ionisation Energy	–
1st Electron Affinity	–
2nd Electron Affinity	–
Enthalpy of Fusion, $\Delta H^\ominus_{Fusion}$	14.4
Enthalpy of Vaporisation, $\Delta H^\ominus_{Vaporisation}$	238.5
Enthalpy of Atomisation, $\Delta H^\ominus_{Atomisation}$	–
	J $K^{-1} mol^{-1}$
Standard Entropy, $S^\ominus$	–

Associated Bond Lengths (nm) & Energies (kJ mol^{-1})	n.a.

Antimony

[Latin, 'stibium'; Greek = 'not alone']

Atomic Number	51
Relative Atomic Mass	121.75

Chemical Symbol	Sb	Group	5	Period	5

Main Source & History

Antimony has been known since ancient times and one of its compounds is mentioned in the Old Testament. It occurs 'free' in small quantities in countries such as Borneo. The main source is stibnite (Sb_2S_3). The element is extracted by reduction with iron or carbon.

Properties

Antimony is a metalloid element with various allotropes. The metal allotrope (ß–antimony) is a silvery–white, brittle, crystalline solid. It is a good conductor of heat. On solidifying from the molten state it expands. Antimony is a relatively unreactive element which does not react with air or acids.

Uses

Antimony is used in semi–conductors and in the manufacture of a wide variety of alloys, including 'Type Metal' and pewter. It is also used in the manufacture of pigments and enamels. Compounds of antimony are used as fluorinating agents (SbF_3), chlorinating agents ($SbCl_5$) and flame retardants in plastics (Sb_2O_3).

Biological Role

Antimony has no natural biological role. It is, however, used in the manufacture of medicines even though the derivatives of antimony are known to be quite toxic.

Melting Point/K	903.9	Boiling Point/K	$1653.0^{ß}$	Density/g cm^{-3}	6.68

Electronic Configuration	[Kr] $4d^{10}$, $5s^2$ $5p^3$

Oxidation State(s)	–3,0,$\underline{+3}$,+4,$\underline{+5}$	Isotopes	^{121}Sb(57.3%),^{123}Sb(42.7%)

S.E.P.	$E^{\ominus}$/Volts
Sb_4O_6|Sb	+0.15

Electronegativity (Pauling)	1.9

Radii	nm
Ionic	0.089 (+3)
Atomic	0.182
Van der Waals	0.220
Covalent	0.141

Enthalpy Data	kJ mol^{-1}
1st Ionisation Energy	834
2nd Ionisation Energy	1595
1st Electron Affinity	–193.0
2nd Electron Affinity	–
Enthalpy of Fusion, $\Delta H^{\ominus}_{Fusion}$	19.8
Enthalpy of Vaporisation, $\Delta H^{\ominus}_{Vaporisation}$	–
Enthalpy of Atomisation, $\Delta H^{\ominus}_{Atomisation}$	262.3
	J $K^{-1}mol^{-1}$
Standard Entropy, $S^{\ominus}$	45.7

Associated Bond Lengths (nm) & Energies (kJ mol^{-1})	Sb–Sb(0.290,299), Sb–Cl(0.233,313), Sb–H(0.171,257), Sb–O(0.200,314)

Argon

[Greek, 'argos' = inactive]

Atomic Number	18
Relative Atomic Mass	39.948

Chemical Symbol	Ar	Group	0 (8)	Period	3

Main Source & History

Discovered by Sir W. Ramsay and Lord Rayleigh in 1894, UK. Argon is the most abundant of all the noble gases in dry air (0.93%). It is extracted from liquid air by fractional distillation.

Properties

It is a colourless, odourless, inert gas. It is a monatomic gas, and there are no interactions between the atoms apart from weak van der Waals forces.

Uses

Argon is used to fill electric light bulbs as well as various types of photo and fluorescent tubes, for metal production (titanium) and as an inert atmosphere in arc welding. It is used in the K–Ar method for dating geological samples.

Biological Role

Argon has no known biological role. It is known be to non–toxic.

Melting Point/K	83.8	Boiling Point/K	87.3	Density/g cm^{-3}	1.38 ^{87K}

Electronic Configuration	[Ne] $3s^2$ $3p^6$

Oxidation State(s)	0	Isotopes	^{40}Ar(99.6%), ^{36}Ar(0.34%)

S.E.P. $E^{\ominus}$/Volts
n.a.

Electronegativity (Pauling)	–

Radii	nm
Ionic	–
Atomic	0.174
Van der Waals	0.192
Covalent	–

Enthalpy Data	kJ mol^{-1}
1st Ionisation Energy	1520
2nd Ionisation Energy	2665
1st Electron Affinity	0.0
2nd Electron Affinity	–
Enthalpy of Fusion, $\Delta H^{\ominus}_{Fusion}$	1.2
Enthalpy of Vaporisation, $\Delta H^{\ominus}_{Vaporisation}$	6.5
Enthalpy of Atomisation, $\Delta H^{\ominus}_{Atomisation}$	0.0
	J $K^{-1}mol^{-1}$
Standard Entropy, $S^{\ominus}$	154.8

Associated Bond Lengths (nm) & Energies (kJ mol^{-1})	n.a.

Arsenic

[Latin, 'arsenikon' = yellow pigment]

Atomic Number	33
Relative Atomic Mass	74.9216

Chemical Symbol	As	Group	5	Period	4

Main Source & History

Initially isolated by Albertus Magnus circa. 1220, Germany. The element occurs native as well as combined in a number of minerals, for example, tealgar (As_2S_2) arsenalite (As_2O_3) and mispickel (FeAsS). The ores are converted to As_2O_3 by roasting in air and this is then reduced with carbon or hydrogen.

Properties

Arsenic is a metalloid element with various allotropes. Grey arsenic is a steel grey, metallic looking substance which is soft and brittle. It is a good conductor of heat. Arsenic burns in oxygen but resists attack by dilute acids, alkalis and water.

Uses

Metallic arsenic is used in semiconductors and in alloys with lead where it has a hardening effect. It is also used in the manufacture of insecticides and weed killers (As_2O_3)

Biological Role

Arsenic is thought to have an essential biological role. It is known to be toxic in small doses and a potential carcinogen. There is a statuatory limit of 1 ppm of this trace element in food, for example ice cream and beer contain 0.5 ppm.

Melting Point/K	1090^{3}	Boiling Point/K	889^{sub}	Density/g cm^{-3}	5.73^{grey}

Electronic Configuration	[Ar] $3d^{10}$, $4s^2$ $4p^3$

Oxidation State(s)	−3,0,±3,±5	Isotopes	^{75}As(100%)

S.E.P.	$E^{\ominus}$/Volts
$HAsO_2$|As	+0.24 (Acid solution)

Electronegativity (Pauling)	2.0

Radii	nm
Ionic	0.047 (+5)
Atomic	0.125
Van der Waals	0.200
Covalent	0.121

Enthalpy Data	kJ mol^{-1}
1st Ionisation Energy	947
2nd Ionisation Energy	1798
1st Electron Affinity	−77.0
2nd Electron Affinity	–
Enthalpy of Fusion, $\Delta H^{\ominus}_{Fusion}$	27.7
Enthalpy of Vaporisation, $\Delta H^{\ominus}_{Vaporisation}$	32.0
Enthalpy of Atomisation, $\Delta H^{\ominus}_{Atomisation}$	302.5
	J $K^{-1}mol^{-1}$
Standard Entropy, $S^{\ominus}$	35.1

Associated Bond Lengths (nm) & Energies (kJ mol^{-1})	As−As(0.244,348), As−Cl(0.216,293), As−H(0.152,245), As−O(0.178,477)

Astatine

[Greek, 'astatos' = unstable]

Atomic Number	85
Relative Atomic Mass	209.9871[1]

Chemical Symbol	At	Group	7	Period	6

Main Source & History

Discovered by D.R. Carson, K.R. MacKenzie and E. Segrè in 1940, California, USA. The only radioactive halogen, traces of which exist with naturally occuring isotopes of uranium, thorium and neptunium. The isotopes ^{209}At and ^{211}At are prepared by bombarding ^{209}Bi with alpha particles.

Properties

Astatine has been found to be as volatile as iodine. Astatine is radioactive with ^{210}At being the longest lived isotope with a $T_{1/2}$ of 8.3 hours.

Uses	Astatine has no known uses.

Biological Role	Astatine has no known biological role. It is hazardous to health due to its radioactive nature. It is thought, like iodine, to accumulate in the thyroid gland.

Melting Point/K	575[2]	Boiling Point/K	610[2]	Density/g cm^{-3}	n.a.

Electronic Configuration	[Xe] $4f^{14}$, $5d^{10}$, $6s^2 6p^5$

Oxidation State(s)	±1,0,+1,+3,+5,+7	Isotopes	^{210}At(0%), ^{211}At(0%)

S.E.P.	$E^\ominus$/Volts
HAtO|At_2	+0.70 (Acid solution)

Electronegativity (Pauling)	2.2

Radii	nm
Ionic	0.227 (−1)
Atomic	–
Van der Waals	–
Covalent	0.145

Enthalpy Data	kJ mol^{-1}
1st Ionisation Energy	930
2nd Ionisation Energy	1600
1st Electron Affinity	−270.0
2nd Electron Affinity	–
Enthalpy of Fusion, $\Delta H^\ominus_{Fusion}$	11.9
Enthalpy of Vaporisation, $\Delta H^\ominus_{Vaporisation}$	45.2
Enthalpy of Atomisation, $\Delta H^\ominus_{Atomisation}$	90.4
	J $K^{-1}mol^{-1}$
Standard Entropy, $S^\ominus$	60.7

Associated Bond Lengths (nm) & Energies (kJ mol^{-1})	At−At(0.290,110)[2]

Barium

[Greek, 'barys' = heavy]

Atomic Number	56
Relative Atomic Mass	137.327

Chemical Symbol	Ba	Group	2	Period	6

Main Source & History

Initially isolated by Sir H. Davy in 1808, London, UK. The principal ore is barytes ($BaSO_4$). It is one of the most difficult metals to extract. The metal is prepared by the electrolysis of fused $BaCl_2$ or by heating BaO with aluminium.

Properties

Barium is a silvery–white metal resembling calcium in appearance. It is a relatively soft metal which oxidises easily in air and is readily attacked by water producing hydrogen gas.

Uses

Barium metal is used in the extraction of americium from AmF_3, where it acts as a reducing agent. It is used to produce barium compounds for the manufacture of paints ($BaCrO_4$), fireworks and glass ($Ba(OH)_2$) as well as for X – ray diagnosis ('Barium Meal' – $BaSO_4$).

Biological Role

Barium has no known biological role. It is a toxic substance.

Melting Point/K	998.0	Boiling Point/K	1910.0	Density/g cm^{-3}	3.51

Electronic Configuration	[Xe] $6s^2$

Oxidation State(s)	–2,0,$\underline{+2}$,+4	Isotopes	^{138}Ba(71.7%), ^{137}Ba(11.3%)

S.E.P.	$E^{\ominus}$/Volts
Ba^{2+}|Ba	–2.92

Electronegativity (Pauling)	0.9

Radii	nm
Ionic	0.136 (+2)
Atomic	0.217
Van der Waals	–
Covalent	0.198

Enthalpy Data	kJ mol^{-1}
1st Ionisation Energy	503
2nd Ionisation Energy	965
1st Electron Affinity	+46.0
2nd Electron Affinity	–
Enthalpy of Fusion, $\Delta H^{\ominus}_{Fusion}$	7.7
Enthalpy of Vaporisation, $\Delta H^{\ominus}_{Vaporisation}$	150.9
Enthalpy of Atomisation, $\Delta H^{\ominus}_{Atomisation}$	180.0
	J $K^{-1}mol^{-1}$
Standard Entropy, $S^{\ominus}$	62.8

Associated Bond Lengths (nm) & Energies (kJ mol^{-1})	n.a.

Berkelium

[After Berkeley, California, USA]

Atomic Number	97
Relative Atomic Mass	247.0703[1]

Chemical Symbol	Bk	Group	–	Period	7 Actinide series

Main Source & History

Discovered by A. Ghiorso, G.T. Seaborg and S.G. Thompson in 1949, Berkley, California, USA. The most stable isotope is produced in an accelerator. ^{247}Bk and ^{249}Bk are produced by neutron bombardment of ^{243}Am or ^{239}Pu. Berkelium is separated by ion exchange. The metal is obtained by lithium reduction of BkF_3.

Properties

It is a silvery, radioactive metal. The most stable isotope ^{247}Bk has a $T_{1/2}$ of 1400 years. It is a relatively electropositive actinide, reacting readily with oxygen, steam and acids.

Uses	Berkelium has no known uses.

Biological Role	Berkelium has no known biological role. It is hazardous to health due to its radioactive nature.

Melting Point/K	1259.0	Boiling Point/K	–	Density/g cm^{-3}	13.25 [293K]

Electronic Configuration	[Rn] $5f^9$, $7s^2$

Oxidation State(s)	0, $\underline{+3}$, +4	Isotopes	^{247}Bk(0%), ^{249}Bk(0%)

S.E.P.	$E^\ominus$/Volts
Bk^{3+}|Bk	– 2.01 Acid solution

Electronegativity (Pauling)	1.3

Radii	nm
Ionic	0.098 (+3)
Atomic	–
Van der Waals	–
Covalent	–

Enthalpy Data	kJ mol^{-1}
1st Ionisation Energy	601
2nd Ionisation Energy	–
1st Electron Affinity	–
2nd Electron Affinity	–
Enthalpy of Fusion, $\Delta H^\ominus_{Fusion}$	–
Enthalpy of Vaporisation, $\Delta H^\ominus_{Vaporisation}$	–
Enthalpy of Atomisation, $\Delta H^\ominus_{Atomisation}$	–
	J $K^{-1}mol^{-1}$
Standard Entropy, $S^\ominus$	–

Associated Bond Lengths (nm) & Energies (kJ mol^{-1})	n.a.

Beryllium

[Greek, 'beryllos' = beryl]

Atomic Number	4
Relative Atomic Mass	9.0122

Chemical Symbol	Be	Group	2	Period	2

Main Source & History

Discovered by N–L. Vauquelin in 1797, and first isolated by A.A. Bussy and F. Wöhler, independantly, in 1828, in France/Germany respectively. The main source is beryl, and beryllium is generally extracted by electrolysis of $BeCl_2$ mixed with NaCl.

Properties

Beryllium is a hard, silvery–white metal of low density and high melting point. It does not react with air, water or steam. Beryllium reacts with sodium hydroxide liberating hydrogen and is rendered passive by nitric acid.

Uses	Beryllium is used in the manufacture of alloys which are used in the building industry and for the manufacture of rocket nose cones. In nuclear reactors it is used as a moderator. Compounds of beryllium are also used in ceramics manufacture (BeO).
Biological Role	Beryllium is toxic and is a known carcinogen. Its compounds have adverse effects on the respiratory system and cause dermatitis.

Melting Point/K	1551.0	Boiling Point/K	3243^3	Density/g cm^{-3}	1.85

Electronic Configuration	[He] $2s^2$

Oxidation State(s)	0,$\pm$2	Isotopes	^{9}Be(100%)

S.E.P.	$E^\ominus$/Volts
Be^{2+}|Be	– 1.85

Electronegativity (Pauling)	1.6

Radii	nm
Ionic	0.027 (+2)
Atomic	0.113
Van der Waals	–
Covalent	0.089

Enthalpy Data	kJ mol^{-1}
1st Ionisation Energy	899
2nd Ionisation Energy	1757
1st Electron Affinity	+66.0
2nd Electron Affinity	–
Enthalpy of Fusion, $\Delta H^\ominus_{Fusion}$	12.0
Enthalpy of Vaporisation, $\Delta H^\ominus_{Vaporisation}$	308.8
Enthalpy of Atomisation, $\Delta H^\ominus_{Atomisation}$	324.3
	J $K^{-1}mol^{-1}$
Standard Entropy, $S^\ominus$	9.5

Associated Bond Lengths (nm) & Energies (kJ mol^{-1})	Be–Be(0.222,–), Be–Cl(0.177,293), Be–H(0.163,226), Be–O(0.133,523)

Bismuth

[German, 'weissmuth' = white matter]

Atomic Number	83
Relative Atomic Mass	208.9804

Chemical Symbol	Bi	Group	5	Period	6

Main Source & History

Isolated by C.F. Geoffroy in 1753, France. It occurs in the 'free' state as well as in sulphide ores such as, bismuth glance (Bi_2S_3). These ores are roasted and the Bi_2O_3 formed is reduced with carbon or hydrogen to give bismuth.

Properties

Bismuth is a silvery–white solid which has a faint reddish tinge. It is a hard, brittle metal with a poor conductivity of heat and electricity. On solidifying from the molten state it expands. Bismuth combines with oxygen on heating and dissolves in concentrated sulphuric acid liberating sulphur dioxide.

Uses	Bismuth is used in the manufacture of low melting point alloys, for example, Woods Metal (m.pt. 333.5K) and in alloys used in the electronics industry. Compounds of bismuth are used in medicines ($BiONO_3.H_2O$), cosmetics (BiOCl) and glazes (Bi_2O_3).
Biological Role	Bismuth has no known biological role. However, it is non–toxic and is used in the manufacture of medicines.

Melting Point/K	544.5	Boiling Point/K	1833.0[2]	Density/g cm^{-3}	9.80

Electronic Configuration	[Xe] $4f^{14}$, $5d^{10}$, $6s^2 6p^3$

Oxidation State(s)	−3,0,$\underline{\pm 3}$,+5	Isotopes	^{209}Bi(100%)

S.E.P.	$E^{\ominus}$/Volts
Bi^{3+}|Bi	+ 0.32

Electronegativity (Pauling)	1.9

Radii	nm
Ionic	0.102 (+3)
Atomic	0.155
Van der Waals	–
Covalent	0.152

Enthalpy Data	kJ mol^{-1}
1st Ionisation Energy	703
2nd Ionisation Energy	1610
1st Electron Affinity	−67.5
2nd Electron Affinity	–
Enthalpy of Fusion, $\Delta H^{\ominus}_{Fusion}$	10.9
Enthalpy of Vaporisation, $\Delta H^{\ominus}_{Vaporisation}$	179.1
Enthalpy of Atomisation, $\Delta H^{\ominus}_{Atomisation}$	207.1
	J $K^{-1}mol^{-1}$
Standard Entropy, $S^{\ominus}$	56.7

Associated Bond Lengths (nm) & Energies (kJ mol^{-1})	Bi–Bi(0.309,200), Bi–Cl(0.248,285), Bi–H(–,194), Bi–O(0.232,339)

Boron

[Persian, 'buraq' = borax]

Atomic Number	5
Relative Atomic Mass	10.811

Chemical Symbol	B	Group	3	Period	2

Main Source & History

Discovered by L.J. Lussac, L.J. Thenard (France) and Sir H. Davy (UK) in 1808. Boron occurs in a number of ores, the most common of which is borax. The element can be obtained by reduction of B_2O_3 with magnesium, in the form of an amorphous powder.

Properties

Boron occurs as two different forms, amorphous and crystalline. Amorphous boron is a dark brown powder which is unreactive to oxygen, acids, alkalis and water. Boron has low electrical conductivity at room temperature which increases with temperature.

Uses	Boron is used extensively as borosilicates in glasses and enamels. It is used in the production of high impact–resistant steel, and other alloys. Due to its ability to absorb neutrons it is also used in control rods for nuclear reactors. It is also mixed with germanium and silicon to improve their conductivities. Compounds of boron are used in medicines ($B(OH)_3$), abrasives (B_4C) and in the manufacture of special glasses (B_2O_3).

Biological Role	Boron has no known biological role in humans but it is essential in plants where it takes part in the calcium cycle.

Melting Point/K	2573.0	Boiling Point/K	2823^{sub}	Density/g cm^{-3}	2.34

Electronic Configuration	[He] $2s^2 2p^1$

Oxidation State(s)	0, $\underline{+3}$	Isotopes	$^{11}B(80\%)$, $^{10}B(20\%)$

S.E.P.	$E^\ominus$/Volts
$B(OH)_3 \| B$	– 0.89

Electronegativity (Pauling)	2.0

Radii	nm
Ionic	0.012 (+3)
Atomic	0.080
Van der Waals	–
Covalent	0.081

Enthalpy Data	kJ mol^{-1}
1st Ionisation Energy	801
2nd Ionisation Energy	2427
1st Electron Affinity	–29.0
2nd Electron Affinity	–
Enthalpy of Fusion, $\Delta H^\ominus_{Fusion}$	22.2
Enthalpy of Vaporisation, $\Delta H^\ominus_{Vaporisation}$	538.9
Enthalpy of Atomisation, $\Delta H^\ominus_{Atomisation}$	562.7
	J $K^{-1}mol^{-1}$
Standard Entropy, $S^\ominus$	5.9

Associated Bond Lengths (nm) & Energies (kJ mol^{-1})	B–B(0.175,335), B–Cl(0.174,444), B–H(0.119,381), B–O(0.136,523), B–F(0.129,644)

Bromine

[Greek, 'bromos' = stench]

Atomic Number	35
Relative Atomic Mass	79.904

Chemical Symbol	Br	Group	7	Period	4

Main Source & History

Discovered by A.J. Balard (France) and C. Löwig (Germany) in 1826. Bromine occurs as bromides in sea water. It is obtained by displacement with chlorine followed by sweeping out with air.

Properties

Bromine is the only liquid halogen at room temperature. It is a deep red, dense, sharp smelling, poisonous liquid. Bromine is very reactive acting as an oxidising agent forming bromides with many elements. It dissolves slightly in water and reacts with alkalis.

Uses

Bromine is used in the manufacture of fuel additives (tetraethyl lead, although to a decreasing extent), flame retardants as well as compounds used in photography (AgBr), disinfectants, medicine (KBr), in herbicides ($C_9H_{13}BrN_2O_2$) as well as oxidising agents ($HBrO_3$).

Biological Role

Bromine has no known biological role due to the toxic nature of Br_2. However, the bromide anion (Br^-) exhibits a specific effect on the central nervous system causing drowsiness and sleep.

Melting Point/K	265.9	Boiling Point/K	331.9	Density/g cm^{-3}	3.12^{293K}

Electronic Configuration	[Ar] $3d^{10}, 4s^2 4p^5$

Oxidation State(s)	$\underline{-1}$,0,+1,+3,+4,+5,+7	Isotopes	^{79}Br(50.7%), ^{81}Br(49.3%)

S.E.P.	$E^\ominus$/Volts
$HBrO \vert Br_2$	+ 1.60 (Acid solution)

Electronegativity (Pauling)	2.8

Radii	nm
Ionic	0.196 (−1)
Atomic	0.114
Van der Waals	0.195
Covalent	0.114

Enthalpy Data	kJ mol^{-1}
1st Ionisation Energy	1140
2nd Ionisation Energy	2100
1st Electron Affinity	−324.5
2nd Electron Affinity	–
Enthalpy of Fusion, $\Delta H^\ominus_{Fusion}$	5.3
Enthalpy of Vaporisation, $\Delta H^\ominus_{Vaporisation}$	15.0
Enthalpy of Atomisation, $\Delta H^\ominus_{Atomisation}$	111.9
	J $K^{-1} mol^{-1}$
Standard Entropy, $S^\ominus$	152.2

Associated Bond Lengths (nm) & Energies (kJ mol^{-1})	Br–Br(0.229,193), Br–H(0.141,366), Br–O(0.160,234), Br–C(0.194,285)

Cadmium

[Latin, 'cadmia' = calomine]

Atomic Number	48
Relative Atomic Mass	112.411

Chemical Symbol	Cd	Group	–	Period	5 2nd series T.M.

Main Source & History

Discovered by F. Stromeyer in 1817, Salzgitter, Sweden. Cadmium is commonly found with zinc in calamine and zinc blende as well as in lead and copper ores. It is formed as a by–product during the production of zinc.

Properties

Cadmium is a silvery–white, lustrous metal. It is soft and is therefore both malleable and ductile. This relatively reactive metal reacts with oxygen and acids but not with alkalis.

Uses

Cadmium is used in the electroplating of iron and steel, as well as in alloys with copper (tramway wires), with aluminium (for casting) and in fusible alloys with tin, bismuth and lead. It is used in the manufacture of Cd–Ni alkaline batteries which are rechargeable. Due to its ability to absorb neutrons it is also used in control rods in nuclear reactors. Compounds of cadmium are used as pigments (CdS).

Biological Role

Cadmium is very toxic and a known carcinogen. It has also been shown to cause malformation of the foetus.

Melting Point/K	594.1	Boiling Point/K	1038.6	Density/g cm^{-3}	8.64

Electronic Configuration	[Kr] $4d^{10}$, $5s^2$

Oxidation State(s)	0,±2	Isotopes	^{114}Cd(28.7%), ^{112}Cd(24.1%)

S.E.P.	$E^{\ominus}$/Volts
Cd^{2+}|Cd	– 0.40 Acid solution

Electronegativity (Pauling)	1.7

Radii	nm
Ionic	0.095 (+2)
Atomic	0.149
Van der Waals	–
Covalent	0.141

Enthalpy Data	kJ mol^{-1}
1st Ionisation Energy	868
2nd Ionisation Energy	1631
1st Electron Affinity	+26.0
2nd Electron Affinity	–
Enthalpy of Fusion, $\Delta H^{\ominus}_{Fusion}$	6.11
Enthalpy of Vaporisation, $\Delta H^{\ominus}_{Vaporisation}$	99.9
Enthalpy of Atomisation, $\Delta H^{\ominus}_{Atomisation}$	112.0
	J $K^{-1}mol^{-1}$
Standard Entropy, $S^{\ominus}$	51.8

Associated Bond Lengths (nm) & Energies (kJ mol^{-1})	n.a.

Caesium

[Latin, 'caesius' = bluish–grey]

Atomic Number	55
Relative Atomic Mass	132.9054

Chemical Symbol	Cs	Group	1	Period	6

Main Source & History

Discovered by R. Bunsen and G.R. Kirchhoff in 1860, Heidelberg, Germany. The main ore is pollucite (hydrated aluminosilicate of caesium). The metal is obtained by calcium metal reduction of the chloride under vacuum.

Properties

It is a very soft, silvery–white metal with a low melting point. Caesium is extremely reactive. It reacts very violently with oxygen, water (liberating hydrogen) and halogens.

Uses

Caesium is used as a reducing agent, as a catalyst for hydrogenation and in photo–electric cells (oxygen–getter). ^{137}Cs is used in deep gamma–ray therapy. ^{133}Cs gives a measure of the standard second.

Biological Role

Caesium has no known biological role.

Melting Point/K	301.6	Boiling Point/K	951.6	Density/g cm^{-3}	1.88

Electronic Configuration	[Xe] $6s^1$

Oxidation State(s)	0,$\pm$1	Isotopes	^{133}Cs(100%)

S.E.P.	$E^\ominus$/Volts
Cs^+|Cs	− 2.92

Electronegativity (Pauling)	0.8

Radii	nm
Ionic	0.169 (+1)
Atomic	0.266
Van der Waals	0.262
Covalent	0.235

Enthalpy Data	kJ mol^{-1}
1st Ionisation Energy	376
2nd Ionisation Energy	2420
1st Electron Affinity	−18.3
2nd Electron Affinity	–
Enthalpy of Fusion, $\Delta H^\ominus_{Fusion}$	2.09
Enthalpy of Vaporisation, $\Delta H^\ominus_{Vaporisation}$	66.5
Enthalpy of Atomisation, $\Delta H^\ominus_{Atomisation}$	76.1
	J K^{-1}mol^{-1}
Standard Entropy, $S^\ominus$	85.2

Associated Bond Lengths (nm) & Energies (kJ mol^{-1})	n.a.

Calcium

[Latin, 'calx' = lime]

Atomic Number	20
Relative Atomic Mass	40.078

Chemical Symbol	Ca	Group	2	Period	4

Main Source & History

Initially isolated by Sir H. Davy in 1808, London, UK. Calcium compounds are widely distributed in nature. They occur as $CaCO_3$ (limestone, marble), $CaSO_4$ (gypsum) as well as silicates and halides. Calcium is mainly extracted by electrolysis of fused $CaCl_2$.

Properties

A relatively soft, silvery–white metal, which is a good conductor of heat and electricity. Calcium reacts rapidly with oxygen, water (liberating hydrogen) and the halogens.

Uses

Calcium is used as a reducing agent in the extraction of thorium, vanadium and zirconium. It is also used in the manufacture of alloys, as a deoxidant in metal castings and for drying organic substances such as absolute alcohol. There are extensive industries based around calcium compounds for example, in the building industry (cement), in agriculture (CaO), in paper and glass manufacture (CaO) and in pharmaceuticals ($CaSO_4$).

Biological Role

Calcium is an essential constituent of body fluids, cells, bones and teeth. The average adult needs to consume about 1 g of calcium per day. It is essential for correct functioning of several bodily processes, for example, the nervous system, muscle contractions and blood clotting. The level of calcium absorbed by our bodies is maintained by vitamin D.

Melting Point/K	1112.0	Boiling Point/K	1757.0	Density/g cm^{-3}	1.54

Electronic Configuration	[Ar] $4s^2$

Oxidation State(s)	0,±2	Isotopes	^{40}Ca(96.9%), ^{44}Ca(2.09%)

S.E.P.	$E^{\ominus}$/Volts
Ca^{2+}|Ca	− 2.87

Electronegativity (Pauling)	1.0

Radii	nm
Ionic	0.100 (+2)
Atomic	0.197
Van der Waals	–
Covalent	0.174

Enthalpy Data	kJ mol^{-1}
1st Ionisation Energy	590
2nd Ionisation Energy	1145
1st Electron Affinity	+186.0
2nd Electron Affinity	–
Enthalpy of Fusion, $\Delta H^{\ominus}_{Fusion}$	9.2
Enthalpy of Vaporisation, $\Delta H^{\ominus}_{Vaporisation}$	150.6
Enthalpy of Atomisation, $\Delta H^{\ominus}_{Atomisation}$	178.2
	J $K^{-1}mol^{-1}$
Standard Entropy, $S^{\ominus}$	41.4

Associated Bond Lengths (nm) & Energies (kJ mol^{-1})	n.a.

Californium

[After California, USA]

Atomic Number	98
Relative Atomic Mass	251.0796[1]

Chemical Symbol	Cf	Group	–	Period	7 Actinide series

Main Source & History

This element was discovered by A. Ghiorso, G.T. Seaborg, K. Street Jr. and S.G. Thompson in 1950, California, USA. The isotopes ^{249}Cf and ^{252}Cf are formed by neutron bombardment of ^{239}Pu. Californium is purified by ion exchange. The metal is isolated by reduction of Cf_2O_3 with lanthanum.

Properties

A silvery–grey, radioactive metal. The most stable isotope ^{251}Cf has a $T_{1/2}$ of 900 years. It is a relatively electropositive actinide, reacting with oxygen, steam and acids.

Uses	Californium has no known uses.

Biological Role	Californium has no known biological role. It is hazardous to health due to its radioactive nature.

Melting Point/K	1173.0	Boiling Point/K	–	Density/g cm^{-3}	15.10^{293K}

Electronic Configuration	[Rn] $5f^{10}$, $7s^2$

Oxidation State(s)	0, +2, $\underline{+3}$, +4	Isotopes	^{249}Cf(0%), ^{251}Cf(0%)

S.E.P.	$E^{\ominus}$/Volts
Cf^{3+}|Cf	– 2.01 (Acid solution)

Electronegativity (Pauling)	1.3

Radii	nm
Ionic	0.098 (+3)
Atomic	–
Van der Waals	–
Covalent	–

Enthalpy Data	kJ mol^{-1}
1st Ionisation Energy	608
2nd Ionisation Energy	–
1st Electron Affinity	–
2nd Electron Affinity	–
Enthalpy of Fusion, $\Delta H^{\ominus}_{Fusion}$	–
Enthalpy of Vaporisation, $\Delta H^{\ominus}_{Vaporisation}$	–
Enthalpy of Atomisation, $\Delta H^{\ominus}_{Atomisation}$	–
	J $K^{-1}mol^{-1}$
Standard Entropy, $S^{\ominus}$	–

Associated Bond Lengths (nm) & Energies (kJ mol^{-1})	n.a.

Carbon

[Latin, 'carbo' = charcoal]

Atomic Number	6
Relative Atomic Mass	12.011

Chemical Symbol	C	Group	4	Period	2

Main Source & History

Carbon has been known since pre–historic times. It is found naturally as the element, as two crystalline allotropes diamond and graphite. Other main sources are hydrocarbons, found in oil and coal and carbonates such as dolomite and limestone.

Properties

Diamond is one of the hardest and most infusible substances known. It is colourless when pure and does not conduct electricity. Graphite is a dull grey, soft solid with high electrical conductivity. Carbon reacts with oxygen at high temperatures, will reduce steam but is not attacked by dilute acids.

Uses

Diamond is used in cutting tools, drills and as a gem. Graphite is used as a lubricant especially for heavy machinery, in pencils, paints, carbon fibres and for carbonising steels. Pure graphite is used as electrodes in the extraction of metals, such as aluminium, and as a moderator in nuclear reactors. ^{14}C is radioactive and is used in carbon dating. Carbon is used for sugar refining (activated charcoal), in printing (carbon black) and in steel making (coke).

Biological Role

Carbon is the basic element of all life. It is present in all organic compounds including DNA. Animals obtain their energy by the oxidation of carbon compounds eaten as food. The biological carbon cycle involves the circulation of carbon (as CO_2) between living organisms and the atmosphere.

Melting Point/K	g:3925sub d:3823	Boiling Point/K	5100	Density/g cm^{-3}	g:2.25 d:3.51

Electronic Configuration	[He] $2s^2$ $2p^2$

Oxidation State(s)	−4,0,+2,±4	Isotopes	^{12}C(98.90%), ^{13}C(1.10%)

S.E.P.	$E^{\ominus}$/Volts
CO|C	+ 0.52 (Acid solution)

Electronegativity (Pauling)	2.5

Radii	nm
Ionic	0.015(+4)
Atomic	0.092
Van der Waals	0.170
Covalent	0.077

Enthalpy Data	kJ mol^{-1}
1st Ionisation Energy	1086
2nd Ionisation Energy	2353
1st Electron Affinity	−120.0
2nd Electron Affinity	–
Enthalpy of Fusion, $\Delta H^{\ominus}_{Fusion}$	105.0
Enthalpy of Vaporisation, $\Delta H^{\ominus}_{Vaporisation}$	716.7 sub
Enthalpy of Atomisation, $\Delta H^{\ominus}_{Atomisation}$	716.7
	J K^{-1} mol^{-1}
Standard Entropy, $S^{\ominus}$	g:5.7 d:2.4

Associated Bond Lengths (nm) & Energies (kJ mol^{-1})	C−C(0.154,348), C=C(0.134,614), C≡C(0.120,839), C−H(0.109,339), C=O(0.123,745), C≡N(0.116,891)

Cerium

[After the asteroid Ceres]

Atomic Number	58
Relative Atomic Mass	140.115

Chemical Symbol	Ce	Group	–	Period	6 Lanthanide series

Main Source & History

Discovered by J.J. Berzelius and W. Hisinger in 1803, Sweden. The most important minerals are cerite, bastnaesite ($CeFCO_3$) and monazite ($(Ce,La,Nd,Pr)PO_4$). Extracted in large quantities from monazite after the thorium has been removed by ion exchange. The metal may be obtained by the reduction of $CeCl_3$ with calcium.

Properties

Cerium, the most abundant of the lanthanides, is a steel–grey, soft metal which is a good conductor of heat and electricity. It tarnishes in air, reacts with water (liberating hydrogen) and also acids and burns in air when ignited.

Uses

Cerium is used in alloys to improve the malleability of cast iron. Cerium–iron alloys are used as 'flints' for lighters. Cerium sulphate is used as the catalyst in 'self–cleaning' ovens and other compounds are used in glass polishing and in gas mantles (CeO_2).

Biological Role

Cerium has no known biological role.

Melting Point/K	1071.0	Boiling Point/K	3699	Density/g cm^{-3}	6.77

Electronic Configuration	[Xe] $4f^2$, $6s^2$

Oxidation State(s)	0,$\underline{+3}$,+4	Isotopes	^{140}Ce(88.48%), ^{142}Ce(11.08%)

S.E.P.	$E^{\ominus}$/Volts
Ce^{3+}|Ce	– 2.33

Electronegativity (Pauling)	1.1

Radii	nm
Ionic	0.107 (+3)
Atomic	0.183
Van der Waals	–
Covalent	0.165

Enthalpy Data	kJ mol^{-1}
1st Ionisation Energy	527
2nd Ionisation Energy	1047
1st Electron Affinity	+50.0[2]
2nd Electron Affinity	–
Enthalpy of Fusion, $\Delta H^{\ominus}_{Fusion}$	8.9
Enthalpy of Vaporisation, $\Delta H^{\ominus}_{Vaporisation}$	398.0
Enthalpy of Atomisation, $\Delta H^{\ominus}_{Atomisation}$	–
	J $K^{-1}mol^{-1}$
Standard Entropy, $S^{\ominus}$	72.0

Associated Bond Lengths (nm) & Energies (kJ mol^{-1})	n.a.

Chlorine

[Greek, 'chloros' = green]

Atomic Number	17
Relative Atomic Mass	35.4527

Chemical Symbol	Cl	Group	7	Period	3

Main Source & History

Initially isolated by C.W. Scheele in 1774, Uppsala, Sweden. It is manufactured almost entirely by the electrolysis of brine. The economics of this process depending on the simultaneous production of sodium hydroxide and hydrogen gas.

Properties

A greenish–yellow, dense, poisonous gas with a choking irritating smell. Chlorine is moderately soluble in water. Solubility increases in organic solvents such as tetrachloromethane. Chlorine is a very reactive element, acting as an oxidising agent forming chlorides with many elements.

Uses

Used in the manufacture of bleaching agents, hydrochloric acid and organo–chlorine solvents (used for degreasing steel and in dry cleaning). Chlorine derivatives are used in water sterilisation (NaOCl), insecticides (CH_3COCH_2Cl), fungicides (CCl_3NO_2), polymer manufacture (PVC), pulp and paper manufacture ($HClO_3$), refrigerants and aerosol propellants (CFC's to a decreasing extent).

Biological Role

The chloride ion (Cl^-), along with the sodium ion (Na^+), is essential in order to cause blood to have an osmotic pressure. It is also involved in the functioning of the central nervous system. The daily requirement is 5–10 mg. In plants it is thought that it may be involved, as Cl^-, in the light stage of photosynthesis. Chlorine gas is very toxic.

Melting Point/K	172.2	Boiling Point/K	239.2	Density/g cm^{-3}	1.56^{239K}

Electronic Configuration	[Ne] $3s^2\ 3p^5$

Oxidation State(s)	<u>–1</u>,0,+1,+3,+4,+5,+6, <u>+7</u>	Isotopes	^{35}Cl(75.77%), ^{37}Cl(24.23%)

S.E.P.	$E^\ominus$/Volts
$^1/_2Cl_2|Cl^-$	+ 1.36

Electronegativity (Pauling)	3.2

Radii	nm
Ionic	0.181 (–1)
Atomic	0.099
Van der Waals	0.180
Covalent	0.099

Enthalpy Data	kJ mol^{-1}
1st Ionisation Energy	1251
2nd Ionisation Energy	2297
1st Electron Affinity	–364.0
2nd Electron Affinity	–
Enthalpy of Fusion, $\Delta H^\ominus_{Fusion}$	3.2
Enthalpy of Vaporisation, $\Delta H^\ominus_{Vaporisation}$	10.2
Enthalpy of Atomisation, $\Delta H^\ominus_{Atomisation}$	121.1
	J $K^{-1}mol^{-1}$
Standard Entropy, $S^\ominus$	165.2

Associated Bond Lengths (nm) & Energies (kJ mol^{-1})	Cl–Cl(0.199,242), Cl– H(0.127,431), Cl–C(0.177,339)

Chromium

[Greek, 'chromos' = colour]

Atomic Number	24
Relative Atomic Mass	51.9961

Chemical Symbol	Cr	Group	–	Period	4 1st series T.M.

Main Source & History

First isolated by N–L. Vauquelin in 1798, Paris, France. The main ore is chromite ($FeCr_2O_4$). Chromium is extracted via the 'Thermite Process', Cr_2O_3 being reduced by aluminium which is obtained by alkaline fusion of the ore and reduction of the Cr(VI) compounds with carbon. The pure metal is obtained by the electrolytic reduction of $CrO_4^{2-}{}_{(aq)}$.

Properties

This transition element is a hard, bluish–white metal very resistant to oxidation. It has the highest melting point in the first transition series. Chromium reacts with oxygen, sulphur and the halogens at high temperatures. It dissolves in sulphuric and hydrochloric acids but is made passive by nitric acid.

Uses	Chromium is used mainly in the production of alloys with iron, eg. stainless steels and for electroplating. Chromium compounds are used as pigments in colouring glass (Cr_2O_3), in leather tanning and as catalysts (Cr_2O_3 is used in the production of methanol) as well as in the manufacture of magnetic tapes (CrO_2). ^{51}Cr is used as a radiation source.
Biological Role	Chromium itself is an essential trace element, but its exact role in the body is not fully known. Chromium compounds are toxic and carcinogenic.

Melting Point/K	2130[2]	Boiling Point/K	2943.0	Density/g cm^{-3}	7.20

Electronic Configuration	[Ar] $3d^5, 4s^1$

Oxidation State(s)	–2, –1, 0, +1, +2, <u>+3</u>, +4, +5, <u>+6</u>	Isotopes	^{52}Cr(83.76%), ^{53}Cr(9.55%)

S.E.P.	$E^⊖$/Volts
Cr^{3+}|Cr	– 0.74 (Acid solution)

Electronegativity (Pauling)	1.7

Radii	nm
Ionic	0.030 (+6)
Atomic	0.125
Van der Waals	–
Covalent	0.118

Enthalpy Data	kJ mol^{-1}
1st Ionisation Energy	653
2nd Ionisation Energy	1592
1st Electron Affinity	+64.6
2nd Electron Affinity	–
Enthalpy of Fusion, $\Delta H^⊖_{Fusion}$	13.8
Enthalpy of Vaporisation, $\Delta H^⊖_{Vaporisation}$	348.8
Enthalpy of Atomisation, $\Delta H^⊖_{Atomisation}$	396.6
	J $K^{-1}mol^{-1}$
Standard Entropy, $S^⊖$	23.5

Associated Bond Lengths (nm) & Energies (kJ mol^{-1})	n.a.

Cobalt

[German, 'kobald' = goblin]

Atomic Number	27
Relative Atomic Mass	58.9332

Chemical Symbol	Co	Group	–	Period	4 1st series T.M.

Main Source & History

First isolated by G. Brandt in 1742, Stockholm, Sweden. The main ores are smaltite ($CoAs_2$) and cobaltite (CoAsS). Extracted by converting the ore into tricobalt tetroxide (Co_3O_4) which is subsequently reduced to the metal by heating with carbon or aluminium.

Properties

The metal is a bluish–white solid which is both malleable and ductile. It is ferromagnetic. Cobalt is a relatively unreactive metal but is attacked by dilute hydrochloric and sulphuric acids and dissolves readily in nitric acid.

Uses

The main use of cobalt is in alloys, eg. stellite (Co,Cr,W), which is extremely hard and used for making high speed cutting tools and valves for the internal combustion engine, and alnico (Al,Ni,Co) used in the manufacture of extremely strong permanent magnets. Cobalt compounds are used in paints (cobalt blue) and as catalysts (for example $Co_2(CO)_8$ is used in the hydroformulation reaction producing an aldehyde or alcohol from an alkene).

Biological Role

Cobalt is an essential trace element. The requirement is less than 1 mg per day. Vitamin B_{12} (cobalamine), a cobalt containing compound, is necessary in the prevention of pernicious anaemia and in the formation of red blood corpuscles. Cobalt can be both toxic and carcinogenic but safety margins are high. In plants it is involved in nitrogen fixation by micro–organisms.

Melting Point/K	1768.0	Boiling Point/K	3143.0	Density/g cm^{-3}	8.90

Electronic Configuration	[Ar] $3d^7$, $4s^2$

Oxidation State(s)	−1,0,+1,$\underline{+2}$,+3,+4,+5	Isotopes	^{59}Co(100%)

S.E.P.	$E^{\ominus}$/Volts
Co^{2+}|Co	− 0.28 (Acid solution)

Electronegativity (Pauling)	1.8

Radii	nm
Ionic	0.078 (+2)
Atomic	0.125
Van der Waals	–
Covalent	0.116

Enthalpy Data	kJ mol^{-1}
1st Ionisation Energy	758
2nd Ionisation Energy	1646
1st Electron Affinity	+70.0
2nd Electron Affinity	–
Enthalpy of Fusion, $\Delta H^{\ominus}_{Fusion}$	15.2
Enthalpy of Vaporisation, $\Delta H^{\ominus}_{Vaporisation}$	382.4
Enthalpy of Atomisation, $\Delta H^{\ominus}_{Atomisation}$	424.7
	J $K^{-1}mol^{-1}$
Standard Entropy, $S^{\ominus}$	30.0

Associated Bond Lengths (nm) & Energies (kJ mol^{-1})	n.a.

Copper

[Latin, 'cuprum' = Cyprus]

Atomic Number	29
Relative Atomic Mass	63.546

Chemical Symbol	Cu	Group	–	Period	4 (1st series T.M.)

Main Source & History

Copper has been known since pre–historic times. Copper occurs native, and also in many copper ores such as cuprite (Cu_2O), malachite ($CuCO_3.Cu(OH)_3$) but the main ore is chalcopyrite ($CuFeS_2$). This ore is concentrated by floatation, roasted in air with SiO_2 to remove iron. The Cu_2S remaining is further roasted to produce copper metal which is purified by electrolysis.

Properties

It is an attractive reddish coloured metal which is very ductile and malleable with thermal and electrical conductivities which are second only to those of silver. Copper is an unreactive metal, being slowly attacked by moist air. It dissolves in oxidising acids.

Uses

Copper metal is primarily used for electrical purposes and pipes (plumbing). It is also used for steam boilers as well as in numerous alloys such as bronze and brass. Copper compounds are used as catalysts (CuCl is used in the manufacture of chlorobenzene), in printing, dyeing ($CuSO_4.5H_2O$) and in the manufacture of paints (CuO) and fungicides ($CuCl_2$).

Biological Role

Copper is an essential trace element, with a probable requirement of 1–3 mg per day. Copper has a role in the formation of haemoglobin. It is a constituent and activator of several enzymes (in plants also) such as ascorbic acid oxidase and lactase. Deficiency of copper leads to anaemia and bone disorders. Excess copper is toxic. There is a recommended limit of 20 ppm in food.

Melting Point/K	1356.4	Boiling Point/K	2840.0	Density/g cm^{-3}	8.92

Electronic Configuration	[Ar] $3d^{10}$, $4s^1$

Oxidation State(s)	0, +1, ±2, +3, +4	Isotopes	^{63}Cu(69.1%), ^{65}Cu(30.9%)

S.E.P.	$E^{\ominus}$/Volts
Cu^{2+}ǀCu	+ 0.34

Electronegativity (Pauling)	1.9

Radii	nm
Ionic	0.072 (+2)
Atomic	0.128
Van der Waals	0.140
Covalent	0.117

Enthalpy Data	kJ mol^{-1}
1st Ionisation Energy	745
2nd Ionisation Energy	1958
1st Electron Affinity	+118.3
2nd Electron Affinity	–
Enthalpy of Fusion, $\Delta H^{\ominus}_{Fusion}$	13.0
Enthalpy of Vaporisation, $\Delta H^{\ominus}_{Vaporisation}$	305.0
Enthalpy of Atomisation, $\Delta H^{\ominus}_{Atomisation}$	338.3
	J $K^{-1}mol^{-1}$
Standard Entropy, $S^{\ominus}$	33.2

Associated Bond Lengths (nm) & Energies (kJ mol^{-1})	n.a.

Curium

[After M. and P Curie]

Atomic Number	96
Relative Atomic Mass	247.0703[1]

Chemical Symbol	Cm	Group	–	Period	7 Actinide series

Main Source & History

Discovered by A. Ghiorso, R.A. James and G.T. Seaberg in 1944, California, USA. ^{242}Cm and ^{244}Cm are obtained in gram quantities by neutron bombardment of ^{239}Pu. The metal is produced by reduction of CmF_3 with barium.

Properties

This element is a silvery, radioactive metal. The most stable isotope ^{247}Cm has a $T_{1/2}$ of 1.6×10^7 years. Curium reacts with oxygen and steam as well as with dilute acids.

Uses	^{244}Cm is used as a neutron target to produce higher actinides.

Biological Role	Curium has no known biological role. It is hazardous to health due to its radioactive nature.

Melting Point/K	1613.0	Boiling Point/K	–	Density/g cm^{-3}	13.51^{293K}

Electronic Configuration	[Rn] $5f^7, 6d^1, 7s^2$

Oxidation State(s)	0, +2, <u>+3</u>, +4	Isotopes	^{247}Cm(0%), ^{248}Cm(0%)

S.E.P.	$E^\ominus$/Volts
Cm^{3+}|Cm	– 2.06 (Acid solution)

Electronegativity (Pauling)	1.3

Radii	nm
Ionic	0.099 (+3)
Atomic	–
Van der Waals	–
Covalent	–

Enthalpy Data	kJ mol^{-1}
1st Ionisation Energy	581
2nd Ionisation Energy	–
1st Electron Affinity	–
2nd Electron Affinity	–
Enthalpy of Fusion, $\Delta H^\ominus_{Fusion}$	–
Enthalpy of Vaporisation, $\Delta H^\ominus_{Vaporisation}$	–
Enthalpy of Atomisation, $\Delta H^\ominus_{Atomisation}$	–
	J $K^{-1}mol^{-1}$
Standard Entropy, $S^\ominus$	–

Associated Bond Lengths (nm) & Energies (kJ mol^{-1})	n.a.

Dysprosium

[Greek, 'dysprositos' = hard to get at]

Atomic Number	66
Relative Atomic Mass	162.50

Chemical Symbol	Dy	Group	–	Period	6 Lanthanide series

Main Source & History

Discovered by Lecoq de Boisbaudran in 1886, Paris, France. It is extracted from monazite and bastnaesite ores with difficulty by ion exchange. The metal may be obtained by reduction of DyF_3 with calcium.

Properties

Dysprosium is a silvery, metallic element. It is hard and reactive and is a good conductor of heat and electricity. Dysprosium reacts rapidly with oxygen, water (liberating hydrogen) and acids.

Uses

Dysprosium metal, as a foil, is used in the measurement of neutron fluxes. Compounds of dysprosium are used in lasers and phosphors (Dy_2O_3).

Biological Role

Dysprosium has no known biological role.

Melting Point/K	1685.0	Boiling Point/K	2835.0	Density/g cm^{-3}	8.56 ^{293K}

Electronic Configuration	[Xe] $4f^{10}$, $6s^2$

Oxidation State(s)	0, +2, <u>+3</u>, +4	Isotopes	^{164}Dy(28.2%), ^{162}Dy(25.53%)

S.E.P.	$E^{\ominus}$/Volts
Dy^{3+}|Dy	– 2.29 (Acid solution)

Electronegativity (Pauling)	1.2

Radii	nm
Ionic	0.091 (+3)
Atomic	0.177
Van der Waals	–
Covalent	0.159

Enthalpy Data	kJ mol^{-1}
1st Ionisation Energy	572
2nd Ionisation Energy	1126
1st Electron Affinity	–
2nd Electron Affinity	–
Enthalpy of Fusion, $\Delta H^{\ominus}_{Fusion}$	17.2
Enthalpy of Vaporisation, $\Delta H^{\ominus}_{Vaporisation}$	293.0
Enthalpy of Atomisation, $\Delta H^{\ominus}_{Atomisation}$	–
	J $K^{-1}mol^{-1}$
Standard Entropy, $S^{\ominus}$	74.8

Associated Bond Lengths (nm) & Energies (kJ mol^{-1})	n.a.

Einsteinium

[After A. Einstein]

Atomic Number	99
Relative Atomic Mass	252.0829[1]

Chemical Symbol	Es	Group	–	Period	7 Actinide series

Main Source & History

Discovered by G.R. Choppin, A. Ghiorso, B.G. Harvey and S.G. Thompson in 1952, in the Pacific amongst the debris created by a thermonuclear explosion. ^{254}Es can be obtained in milligram quantities by neutron bombardment of Am, Pu and Cm. It can be purified by ion exchange.

Properties

This transuranic element is a silvery radioactive metal. The most stable isotope ^{254}Es has a $T_{1/2}$ of 207 days. Einsteinium is a relatively reactive element being attacked by oxygen, steam and acids.

Uses	Einsteinium has no known uses.

Biological Role	Einsteinium has no known biological role. It is hazardous to health due to its radioactive nature.

Melting Point/K	–	Boiling Point/K	–	Density/g cm^{-3}	–

Electronic Configuration	[Rn] $5f^{11}$, $7s^2$

Oxidation State(s)	0, +2, <u>+3</u>	Isotopes	^{254}Es(0%), ^{253}Es(0%)

S.E.P.	$E^{\ominus}$/Volts
Es^{3+}|Es	– 1.98 (Acid solution)

Electronegativity (Pauling)	1.3

Radii	nm
Ionic	0.098 (+3)
Atomic	–
Van der Waals	–
Covalent	–

Enthalpy Data	kJ mol^{-1}
1st Ionisation Energy	619
2nd Ionisation Energy	–
1st Electron Affinity	+50.0[2]
2nd Electron Affinity	–
Enthalpy of Fusion, $\Delta H^{\ominus}_{Fusion}$	–
Enthalpy of Vaporisation, $\Delta H^{\ominus}_{Vaporisation}$	–
Enthalpy of Atomisation, $\Delta H^{\ominus}_{Atomisation}$	–
	J K^{-1} mol^{-1}
Standard Entropy, $S^{\ominus}$	–

Associated Bond Lengths (nm) & Energies (kJ mol^{-1})	n.a.

Erbium

[After Ytterby, Sweden]

Atomic Number	68
Relative Atomic Mass	167.26

Chemical Symbol	Er	Group	–	Period	6 Lanthanide series

Main Source & History

Discovered by C.G. Mosander in 1843, Stockholm, Sweden. The main ores are monazite and bastnaesite from which erbium is extracted with difficulty by ion exchange. The metal may be obtained by the reduction of ErF_3 with calcium.

Properties

Erbium is a silvery–white metal which is a good conductor of heat and electricity. It reacts directly with water (liberating hydrogen) and acids. Erbium also reacts with air.

Uses

Erbium is used as an additive to improve the malleability of vanadium as well as in infrared absorbing glass. Erbium(III) oxide is used as a pigment in glass.

Biological Role

Erbium has no known biological role.

Melting Point/K	1802.0	Boiling Point/K	3136.0	Density/g cm^{-3}	9.05

Electronic Configuration	[Xe] $4f^{12}$, $6s^2$

Oxidation State(s)	0,$\underline{\pm 3}$	Isotopes	^{166}Er(33.41%), ^{168}Er(27.07%)

S.E.P.	$E^\ominus$/Volts
Er^{3+}|Er	– 2.32 Acid solution

Electronegativity (Pauling)	1.2

Radii	nm
Ionic	0.089 (+3)
Atomic	0.173
Van der Waals	–
Covalent	0.157

Enthalpy Data	kJ mol^{-1}
1st Ionisation Energy	589
2nd Ionisation Energy	1151
1st Electron Affinity	$+50.0^2$
2nd Electron Affinity	–
Enthalpy of Fusion, $\Delta H^\ominus_{Fusion}$	17.2
Enthalpy of Vaporisation, $\Delta H^\ominus_{Vaporisation}$	280.0
Enthalpy of Atomisation, $\Delta H^\ominus_{Atomisation}$	–
	J $K^{-1}mol^{-1}$
Standard Entropy, $S^\ominus$	73.2

Associated Bond Lengths (nm) & Energies (kJ mol^{-1})	n.a.

Europium

[After Europe]

Atomic Number	63
Relative Atomic Mass	151.965

Chemical Symbol	Eu	Group	–	Period	6 Lanthanide series

Main Source & History

Discovered by E.A. Demarçay in 1901, Paris, France. The main ores are monazite and bastnaesite from which europium is extracted with difficulty, due to its extreme rareness, by ion exchange. The metal may be obtained by the reduction of europium(III) oxide.

Properties

Europium is a silvery–white element. It is a soft, reactive metal which is a good conductor of heat and electricity. Europium reacts readily with oxygen and water from which hydrogen is liberated. It also reacts with acids.

Uses

Europium is used as a neutron absorber in nuclear reactors. Compounds of europium are used to produce red phosphors for colour televisions.

Biological Role

Europium has no known biological role.

Melting Point/K	1095.0	Boiling Point/K	1870.0	Density/g cm^{-3}	5.25

Electronic Configuration	[Xe] $4f^7$, $6s^2$

Oxidation State(s)	0, +2, <u>+3</u>	Isotopes	^{153}Eu(52.23%), ^{151}Er(47.77%)

S.E.P.	$E^{\ominus}$/Volts
$Eu^{3+}\|Eu$	– 1.99 Acid solution

Electronegativity (Pauling)	1.1

Radii	nm
Ionic	0.095 (+3)
Atomic	0.200
Van der Waals	–
Covalent	0.185

Enthalpy Data	kJ mol^{-1}
1st Ionisation Energy	547
2nd Ionisation Energy	1085
1st Electron Affinity	+50.0[2]
2nd Electron Affinity	–
Enthalpy of Fusion, $\Delta H^{\ominus}_{Fusion}$	10.5
Enthalpy of Vaporisation, $\Delta H^{\ominus}_{Vaporisation}$	176.0
Enthalpy of Atomisation, $\Delta H^{\ominus}_{Atomisation}$	–
	J $K^{-1}mol^{-1}$
Standard Entropy, $S^{\ominus}$	77.8

Associated Bond Lengths (nm) & Energies (kJ mol^{-1})	n.a.

Fermium

[After E. Fermi]

Atomic Number	100
Relative Atomic Mass	257.0951[1]

Chemical Symbol	Fm	Group	–	Period	7 Actinide series

Main Source & History

Discovered by G.R. Choppin, A. Ghiorso, B.G. Harvey and S.G. Thompson in 1952, in the Pacific amongst the debris created by a thermonuclear explosion. ^{253}Fm can be obtained in microgram quantities by neutron bombardment of ^{239}Pu. It can be purified by ion exchange.

Properties

This transuranic element is a silvery, very radioactive metal and has not yet been prepared in sufficient quantities for chemical testing. The most stable isotope ^{257}Fm has a $T_{1/2}$ of 80 days.

Uses

Fermium has no known uses.

Biological Role

Fermium has no known biological role. It is hazardous to health due to its radioactive nature.

Melting Point/K	–	Boiling Point/K	–	Density/g cm^{-3}	–

Electronic Configuration	[Rn] $5f^{12}$, $7s^2$

Oxidation State(s)	0, +2, $\underline{+3}$	Isotopes	^{255}Fm(0%), ^{257}Fm(0%)

S.E.P.	$E^{\ominus}$/Volts
Fm^{3+}|Fm	– 1.96 Acid solution

Electronegativity (Pauling)	1.3

Radii	nm
Ionic	0.097 (+3)
Atomic	–
Van der Waals	–
Covalent	–

Enthalpy Data	kJ mol^{-1}
1st Ionisation Energy	627
2nd Ionisation Energy	–
1st Electron Affinity	–
2nd Electron Affinity	–
Enthalpy of Fusion, $\Delta H^{\ominus}_{Fusion}$	–
Enthalpy of Vaporisation, $\Delta H^{\ominus}_{Vaporisation}$	–
Enthalpy of Atomisation, $\Delta H^{\ominus}_{Atomisation}$	–
	J $K^{-1}mol^{-1}$
Standard Entropy, $S^{\ominus}$	–

Associated Bond Lengths (nm) & Energies (kJ mol^{-1})	n.a.

Fluorine

[Latin, 'fluo' = flux]

Atomic Number	9
Relative Atomic Mass	18.9984

Chemical Symbol	F	Group	7	Period	2

Main Source & History

First isolated by H. Moissan in 1886, Paris, France. There are a large number of minerals containing fluorine, eg. cryolite (Na_3AlF_6) and fluorite (CaF_2). The element is obtained by the electrolysis of a HF–KF electrolyte mixture. Fluorine is liberated at the carbon anode. It is purified by passing the gas over sodium fluoride pellets.

Properties

Fluorine is a pale yellow gas (F_2) with an irritating smell. It is extremely poisonous and the most chemically reactive non–metal. Fluorine is extremely oxidising and forms fluorides with most other elements.

Uses

Fluorine is used to make a variety of very useful compounds. These are used in the manufacture of polymers (Teflon), refrigerants and aerosol propellants (CFC's to a decreasing extent), toothpaste (Na_2PO_3F) and in water treatment (KF). They are also used in the separation of uranium isotopes (UF_6).

Biological Role

Fluorine is an essential trace element which is found in bones and teeth. The requirement is 1–2 mg per day. Fluoride ions (F^-) in low concentrations in drinking water can effectively prevent tooth decay in children. At high concentrations the ion is toxic.

Melting Point/K	53.5	Boiling Point/K	85.0	Density/g cm^{-3}	1.51^{85K}

Electronic Configuration	[He] $2s^2$ $2p^5$

Oxidation State(s)	–1,0	Isotopes	^{19}F(100%)

S.E.P.	$E^\ominus$/Volts
$^1/_2F_2\|F^-$	+ 2.87

Electronegativity (Pauling)	4.0

Radii	nm
Ionic	0.133 (–1)
Atomic	0.071
Van der Waals	0.135
Covalent	0.064

Enthalpy Data	kJ mol^{-1}
1st Ionisation Energy	1681
2nd Ionisation Energy	3374
1st Electron Affinity	–348.0
2nd Electron Affinity	–
Enthalpy of Fusion, $\Delta H^\ominus_{Fusion}$	0.26
Enthalpy of Vaporisation, $\Delta H^\ominus_{Vaporisation}$	3.3
Enthalpy of Atomisation, $\Delta H^\ominus_{Atomisation}$	79.1
	J $K^{-1}mol^{-1}$
Standard Entropy, $S^\ominus$	202.8

Associated Bond Lengths (nm) & Energies (kJmol^{-1})	F–F(0.142,158), F–H(0.092,562), F–C(0.138,467)

Francium

[After France]

Atomic Number	87
Relative Atomic Mass	223.0197[1]

Chemical Symbol	Fr	Group	1	Period	7

Main Source & History

Discovered by Mlle. M. Perey in 1939, Paris, France. The element is obtained in very small quantities as part of a number of natural decay series.

Properties

Francium is a short lived radioactive metallic element. The most stable isotope ^{223}Fr has a $T_{1/2}$ of 22 minutes. It is a member of the alkali metals.

Uses

Francium has no known uses.

Biological Role

Francium has no known biological role. It is hazardous to health due to its radioactive nature.

Melting Point/K	300.0[2]	Boiling Point/K	950.0[2]	Density/g cm^{-3}	–

Electronic Configuration	[Rn] $7s^1$

Oxidation State(s)	0,$\pm$1	Isotopes	^{223}Fr(trace), ^{212}Fr(0%)

S.E.P.	$E^\ominus$/Volts
Fr^+|Fr	– 2.9[2]

Electronegativity (Pauling)	0.7

Radii	nm
Ionic	0.180 (+1)
Atomic	0.270[2]
Van der Waals	–
Covalent	–

Enthalpy Data	kJ mol^{-1}
1st Ionisation Energy	400
2nd Ionisation Energy	2100[2]
1st Electron Affinity	–44[4]
2nd Electron Affinity	–
Enthalpy of Fusion, $\Delta H^\ominus_{Fusion}$	2.1[2]
Enthalpy of Vaporisation, $\Delta H^\ominus_{Vaporisation}$	63.6[2]
Enthalpy of Atomisation, $\Delta H^\ominus_{Atomisation}$	72.8[2]
	J $K^{-1}mol^{-1}$
Standard Entropy, $S^\ominus$	95.4

Associated Bond Lengths (nm) & Energies (kJ mol^{-1})	n.a.

Gadolinium

[After J. Gadolin]

Atomic Number	64
Relative Atomic Mass	157.25

Chemical Symbol	Gd	Group	–	Period	6 Lanthanide series

Main Source & History

Discovered by J–C. G. de Marignac in 1880, Geneva, Switzerland. The main ores are monazite and bastnaesite from which gadolinium is extracted with difficulty by ion exchange. The metal may be obtained by the reduction of $GdCl_3$ with calcium.

Properties

Gadolinium is a silvery–white metal which is a good conductor of heat and electricity. It is a relatively unreactive element reacting only slowly with oxygen and water (liberating hydrogen). It also reacts with acids.

Uses

Gadolinium is used in the manufacture of alloys which are used in the electronics industry. Compounds of gadolinium are used as catalysts in the addition polymerisation of alkenes.

Biological Role

Gadolinium has no known biological role.

Melting Point/K	1586.0	Boiling Point/K	3540.0	Density/g cm^{-3}	7.90

Electronic Configuration	[Xe] $4f^7, 5d^1, 6s^2$

Oxidation State(s)	0, +2, ±3	Isotopes	^{158}Gd(24.9%), ^{160}Gd(21.8%)

S.E.P.	$E^{\ominus}$/Volts
Gd^{3+}|Gd	– 2.29 Acid solution

Electronegativity (Pauling)	1.2

Radii	nm
Ionic	0.097 (+3)
Atomic	0.179
Van der Waals	–
Covalent	0.161

Enthalpy Data	kJ mol^{-1}
1st Ionisation Energy	593
2nd Ionisation Energy	1167
1st Electron Affinity	+50.0[2]
2nd Electron Affinity	–
Enthalpy of Fusion, $\Delta H^{\ominus}_{Fusion}$	15.5
Enthalpy of Vaporisation, $\Delta H^{\ominus}_{Vaporisation}$	301.0
Enthalpy of Atomisation, $\Delta H^{\ominus}_{Atomisation}$	–
	J K^{-1} mol^{-1}
Standard Entropy, $S^{\ominus}$	68.1

Associated Bond Lengths (nm) & Energies (kJ mol^{-1})	n.a.

Gallium

[Latin, 'Gallia' = France]

Atomic Number	31
Relative Atomic Mass	69.723

Chemical Symbol	Ga	Group	3	Period	4

Main Source & History

Discovered by L. de Boisbaudran in 1875, Paris, France. It is found in the mineral germanite. Gallium can be obtained by electrolysis of an alkaline solution of one of its salts.

Properties

Gallium is a soft, silvery–white metal. It is a liquid over an extremely large range of temperature. Gallium reacts with acids, water and alkalis (liberating hydrogen) as well as with air.

Uses	Gallium is used in high temperature thermometers, in doping semiconductors, phosphors and light emitting diodes (as GaAs and GaP).

Biological Role	Gallium has no known biological role.

Melting Point/K	302.9	Boiling Point/K	2676.0	Density/g cm^{-3}	5.91^{293K}

Electronic Configuration	[Ar] $3d^{10}, 4s^2 4p^1$

Oxidation State(s)	0, +1, +2, $\underline{+3}$	Isotopes	^{69}Ga(60.2%), ^{71}Ga(39.8%)

S.E.P.	$E^{\ominus}$/Volts
Ga^{3+}|Ga	− 0.53 (Acid solution)

Electronegativity (Pauling)	1.6

Radii	nm
Ionic	0.062 (+3)
Atomic	0.122
Van der Waals	0.190
Covalent	0.125

Enthalpy Data	kJ mol^{-1}
1st Ionisation Energy	579
2nd Ionisation Energy	1979
1st Electron Affinity	−36[4]
2nd Electron Affinity	–
Enthalpy of Fusion, $\Delta H^{\ominus}_{Fusion}$	5.6
Enthalpy of Vaporisation, $\Delta H^{\ominus}_{Vaporisation}$	256.0
Enthalpy of Atomisation, $\Delta H^{\ominus}_{Atomisation}$	277.0
	J $K^{-1} mol^{-1}$
Standard Entropy, $S^{\ominus}$	40.9

Associated Bond Lengths (nm) & Energies (kJ mol^{-1})	n.a.

Germanium

[Latin, 'Germania' = Germany]

Atomic Number	32
Relative Atomic Mass	72.61

Chemical Symbol	Ge	Group	4	Period	4

Main Source & History

Discovered by C.A. Winkler in 1886, Germany. The main ore is germanite. Germanium in the ore is converted to $GeCl_4$ and hydrolysed to GeO_2. This oxide is then reduced with carbon or hydrogen. Germanium can be purified by zone refining.

Properties

Germanium is a metalloid element. It is a silvery–white, hard and brittle element with a structure similar to that of diamond. It is a relatively unreactive element but reacts with nitric acid and alkalis.

Uses	Germanium is used as a semiconductor, in phosphors and in the manufacture of alloys which expand on freezing. It is also used in the production of special glasses.
Biological Role	Germanium has no known biological role.

Melting Point/K	1210.4	Boiling Point/K	3103.0	Density/g cm^{-3}	5.35

Electronic Configuration	[Ar] $3d^{10},4s^2\ 4p^2$

Oxidation State(s)	$-4, 0, +2, \underline{+4}$	Isotopes	^{74}Ge(36.5%), ^{72}Ge(27.4%)

S.E.P.	$E^{\ominus}$/Volts
Ge^{2+}|Ge	− 0.25 (Acid solution)

Electronegativity (Pauling)	1.8

Radii	nm
Ionic	0.054 (+4)
Atomic	0.123
Van der Waals	–
Covalent	0.122

Enthalpy Data	kJ mol^{-1}
1st Ionisation Energy	762
2nd Ionisation Energy	1537
1st Electron Affinity	−116.0
2nd Electron Affinity	–
Enthalpy of Fusion, $\Delta H^{\ominus}_{Fusion}$	32.0
Enthalpy of Vaporisation, $\Delta H^{\ominus}_{Vaporisation}$	330.0
Enthalpy of Atomisation, $\Delta H^{\ominus}_{Atomisation}$	376.6
	J $K^{-1}mol^{-1}$
Standard Entropy, $S^{\ominus}$	31.1

Associated Bond Lengths (nm) & Energies (kJ mol^{-1})	Ge–Ge(0.241,163), Ge–H(0.153,288), Ge–O(0.165,363), Ge–Cl(0.210,340)

Gold

[Anglo–Saxon, gold]

Atomic Number	79
Relative Atomic Mass	196.9665

Chemical Symbol	Au	Group	–	Period	6 3rd series T.M.

Main Source & History

Gold has been known since pre–historic times. It occurs mainly as the free metal in quartz veins, from which it is extracted by the cyanide process, and in alluvial gravels from which it can be separated by mechanical methods. It is refined electrolytically.

Properties

Gold is a bright yellow, metallic element. It is extremely ductile and malleable as well as an extremely good conductor of both heat and electricity. Gold is not attacked by air, water or most acids. However, it will react with aqua regia and H_2SeO_4.

Uses

Gold is primarily used in monetary systems. It is also used in jewellrey and the electronic industry as well as in glass as an infra–red reflector. The isotope ^{198}Au is used as a gamma radiation source.

Biological Role

Gold has no known biological role. Compounds of gold are used in the treatment of rheumatoid arthritis.

Melting Point/K	1337.6	Boiling Point/K	3080.0	Density/g cm^{-3}	19.29

Electronic Configuration	[Xe] $4f^{14}$, $5d^{10}$, $6s^{1}$

Oxidation State(s)	– 1,0,+1,+2,$\underline{+3}$,+5	Isotopes	^{197}Au(100%)

S.E.P.	$E^{\ominus}$/Volts
Au^{3+}|Au	+ 1.50 Acid solution

Electronegativity (Pauling)	2.4

Radii	nm
Ionic	$0.091^{2}(+3)$
Atomic	0.144
Van der Waals	–
Covalent	0.134

Enthalpy Data	kJ mol^{-1}
1st Ionisation Energy	890
2nd Ionisation Energy	1980
1st Electron Affinity	+222.7
2nd Electron Affinity	–
Enthalpy of Fusion, $\Delta H^{\ominus}_{Fusion}$	12.7
Enthalpy of Vaporisation, $\Delta H^{\ominus}_{Vaporisation}$	343.1
Enthalpy of Atomisation, $\Delta H^{\ominus}_{Atomisation}$	369.0
	J $K^{-1}mol^{-1}$
Standard Entropy, $S^{\ominus}$	47.4

Associated Bond Lengths (nm) & Energies (kJ mol^{-1})	n.a.

Hafnium

[Latin, 'Hafnia' = Copenhagen]

Atomic Number	72
Relative Atomic Mass	178.49

Chemical Symbol	Hf	Group	–	Period	6 3rd series T.M.

Main Source & History

Discovered by D. Coster and G.C. von Hevesey in 1923, Copenhagen, Denmark. It occurs in the zirconium containing minerals baddeleyite and zircon from which it is separated by ion exchange.

Properties

Hafnium is a silvery–grey metal. It is lustrous, hard and ductile. Hafnium is a relatively unreactive element. It resists corrosion and is unaffected by common acids and alkalis.

Uses

Hafnium is used in the production of alloys with tantalum and tungsten (in filaments). It is a strong absorber of neutrons and is used to make control rods in nuclear reactors.

Biological Role

Hafnium has no known biological role.

Melting Point/K	2500[2]	Boiling Point/K	4875.0	Density/g cm^{-3}	13.31

Electronic Configuration	[Xe] $4f^{14}$, $5d^2$, $6s^2$

Oxidation State(s)	0, +3, <u>+4</u>	Isotopes	^{180}Hf(35.2%), ^{178}Hf(27.1%)

S.E.P.	$E^\ominus$/Volts
Hf^{4+}|Hf	– 1.70

Electronegativity (Pauling)	1.3

Radii	nm
Ionic	0.081 (+4)
Atomic	0.156
Van der Waals	–
Covalent	0.144

Enthalpy Data	kJ mol^{-1}
1st Ionisation Energy	642[2]
2nd Ionisation Energy	1440
1st Electron Affinity	+61.0
2nd Electron Affinity	–
Enthalpy of Fusion, $\Delta H^\ominus_{Fusion}$	21.8
Enthalpy of Vaporisation, $\Delta H^\ominus_{Vaporisation}$	648.0
Enthalpy of Atomisation, $\Delta H^\ominus_{Atomisation}$	669.0[2]
	J $K^{-1}mol^{-1}$
Standard Entropy, $S^\ominus$	43.6

Associated Bond Lengths (nm) & Energies (kJ mol^{-1})	n.a.

Helium

[Greek, 'helios' = sun]

Atomic Number	2
Relative Atomic Mass	4.0026

Chemical Symbol	He	Group	0 (8)	Period	1

Main Source & History

First isolated by Sir W. Ramsay (UK) and by P.T. Cleve and N.A. Langlet (Sweden) in 1894. It is found in the atmosphere and in natural gas. It is extracted from liquid air by fractional distillation.

Properties

It is a colourless, odourless, inert gas. It is a monatomic gas, and there are no interactions between the atoms apart from weak van der Waals forces.

Uses

Helium is used to provide an inert shield for welding, as a coolant in nuclear reactors and, with 20% oxygen, as a breathing gas used by divers. Helium is also used to inflate the tyres of large aircraft, for filling balloons and in lasers. As liquid helium it is used in low temperature research.

Biological Role

Helium has no known biological role. It is known be to non–toxic.

Melting Point/K	1.0^{26atm}	Boiling Point/K	4.1	Density/g cm^{-3}	0.15^{3K}

Electronic Configuration	$1s^2$

Oxidation State(s)	0	Isotopes	$^4He(100\%)$, $^3He(trace)$

S.E.P. $E^{\ominus}$/Volts
n.a.

Electronegativity (Pauling)	–

Radii	nm
Ionic	–
Atomic	0.128
Van der Waals	0.180
Covalent	–

Enthalpy Data	kJ mol^{-1}
1st Ionisation Energy	2372
2nd Ionisation Energy	5250
1st Electron Affinity	0
2nd Electron Affinity	–
Enthalpy of Fusion, $\Delta H^{\ominus}_{Fusion}$	0.02
Enthalpy of Vaporisation, $\Delta H^{\ominus}_{Vaporisation}$	0.08
Enthalpy of Atomisation, $\Delta H^{\ominus}_{Atomisation}$	0.0
	J $K^{-1}mol^{-1}$
Standard Entropy, $S^{\ominus}$	126.2

Associated Bond Lengths (nm) & Energies (kJ mol^{-1})	n.a.

Holmium

[Latin, 'Holmia' = Stockholm]

Atomic Number	67
Relative Atomic Mass	164.9303

Chemical Symbol	Ho	Group	–	Period	6 Lanthanide series

Main Source & History

Discovered by P.T. Cleve (Sweden) and J–L. Soret (Switzerland) in 1878. The main ores are monazite and bastnaesite from which holmium is extracted with difficulty by ion exchange. The metal may be obtained by reduction of HoF_3 with calcium.

Properties

Holmium is a silvery–white metal which is a good conductor of heat and electricity. Holmium is relatively unreactive. It reacts only slowly with oxygen and water, however, it does react with acids.

Uses	Holmium has no known uses.

Biological Role	Holmium has no known biological role.

Melting Point/K	1747.0	Boiling Point/K	2973.0	Density/g cm^{-3}	8.79

Electronic Configuration	[Xe] $4f^{11}$, $6s^2$

Oxidation State(s)	0,$\underline{+3}$	Isotopes	^{165}Ho(100%)

S.E.P.	$E^{\ominus}$/Volts
Ho^{3+}|Ho	– 2.33 Acid solution

Electronegativity (Pauling)	1.2

Radii	nm
Ionic	0.090 (+3)
Atomic	0.174
Van der Waals	–
Covalent	0.158

Enthalpy Data	kJ mol^{-1}
1st Ionisation Energy	581
2nd Ionisation Energy	1139
1st Electron Affinity	+50.0[2]
2nd Electron Affinity	–
Enthalpy of Fusion, $\Delta H^{\ominus}_{Fusion}$	17.2
Enthalpy of Vaporisation, $\Delta H^{\ominus}_{Vaporisation}$	303.0
Enthalpy of Atomisation, $\Delta H^{\ominus}_{Atomisation}$	–
	J $K^{-1}mol^{-1}$
Standard Entropy, $S^{\ominus}$	75.3

Associated Bond Lengths & Energies	n.a.

Hydrogen

[Greek, 'hydro genes' = water producer]

Atomic Number	1
Relative Atomic Mass	1.0079

Chemical Symbol	H	Group	–	Period	1

Main Source & History

It has been known since the sixteenth century, however, it was first identified by H. Cavendish in 1766, London, UK. Hydrogen is obtained in large quantities by the catalytic reforming of methane.

Properties

Hydrogen is a colourless, tasteless and odourless gas with the lowest density of all elements. It is relatively insoluble in water but is extremely flammable. Hydrogen is used extensively as a reducing agent.

Uses

Hydrogen is used in the manufacture of ammonia, methanol, margarines, synthetic oil, hydrogen chloride gas and hydrochloric acid. It is also used in the removal of sulphur from oil as well as in the oxy–hydrogen burner and as a reducing agent in the extraction of many metals, eg. tungsten. Liquid hydrogen is used for low temperature research and as a coolant for generators in power stations. There are numerous uses for the enormous range of compounds containing hydrogen.

Biological Role

Hydrogen is present in virtually all the essential constituent compounds necessary for life, for example it is found in DNA.

Melting Point/K	14.0	Boiling Point/K	20.0	Density/g cm^{-3}	0.07^{20K}

Electronic Configuration	$1s^1$

Oxidation State(s)	$-1, 0, \underline{+1}$	Isotopes	^{1}H(99.985%), ^{2}H(0.015%)

S.E.P.	$E^{\ominus}$/Volts
$2H^+ \vert H_2$	0.00 (Acid solution)

Electronegativity (Pauling)	2.1

Radii	nm
Ionic	0.208 (−1)
Atomic	0.037
Van der Waals	0.120
Covalent	0.030

Enthalpy Data	kJ mol^{-1}
1st Ionisation Energy	1312
2nd Ionisation Energy	–
1st Electron Affinity	−72.8
2nd Electron Affinity	–
Enthalpy of Fusion, $\Delta H^{\ominus}_{Fusion}$	0.06
Enthalpy of Vaporisation, $\Delta H^{\ominus}_{Vaporisation}$	0.45
Enthalpy of Atomisation, $\Delta H^{\ominus}_{Atomisation}$	218.0
	J $K^{-1}mol^{-1}$
Standard Entropy, $S^{\ominus}$	131.0

Associated Bond Lengths (nm) & Energies (kJ mol^{-1})	H–H(0.074,454), H–Cl(0.127,431), H–O(0.096,464), H–F(0.092,562), H–C(0.109,339)

Indium

[After a colour in its line spectrum]

Atomic Number	49
Relative Atomic Mass	114.82

Chemical Symbol	In	Group	3	Period	5

Main Source & History

Discovered by F. Reich and H. Richter in 1863, Germany. It occurs in zinc blende and in cylindrite. The metal is precipitated from solution by zinc and then purified electrolytically.

Properties

Indium is a soft, silvery–white metal with a low melting point. It is a liquid over a large temperature range. Indium is a fairly reactive metal combining with many non–metals, eg. sulphur and the halogens and dissolves in acids. However, indium is stable towards air and water.

Uses	Indium is used in electronic components and semiconductors (InAs and InSb). It is used in low melting point alloys, such as dental alloys, in electroplating and in electric motors.

Biological Role	Indium has no known biological role. Its compounds are toxic and the presence of indium in the body can cause malformation of the foetus.

Melting Point/K	429.3	Boiling Point/K	2353.0	Density/g cm^{-3}	7.31

Electronic Configuration	[Kr] $4d^{10}$, $5s^2$ $5p^1$

Oxidation State(s)	0, +1, +2, <u>+3</u>	Isotopes	^{115}In(95.7%), ^{113}In(4.3%)

S.E.P.	$E^{\ominus}$/Volts
In^{3+}|In	−0.34 (Acid solution)

Electronegativity (Pauling)	1.8

Radii	nm
Ionic	0.081 (+3)
Atomic	0.163
Van der Waals	–
Covalent	0.150

Enthalpy Data	kJ mol^{-1}
1st Ionisation Energy	558
2nd Ionisation Energy	1821
1st Electron Affinity	−29.0
2nd Electron Affinity	–
Enthalpy of Fusion, $\Delta H^{\ominus}_{Fusion}$	3.3
Enthalpy of Vaporisation, $\Delta H^{\ominus}_{Vaporisation}$	226.4
Enthalpy of Atomisation, $\Delta H^{\ominus}_{Atomisation}$	243.3
	J $K^{-1}mol^{-1}$
Standard Entropy, $S^{\ominus}$	57.8

Associated Bond Lengths (nm) & Energies (kJ mol^{-1})	In–In(0.325,85^2), In–H(0.185,243), In–O(0.213,109), In–Cl(0.240,439)

Iodine

[Greek, 'iodes' = violet]

Atomic Number	53
Relative Atomic Mass	126.9045

Chemical Symbol	I	Group	7	Period	5

Main Source & History

Discovered by B. Curtois in 1811, Paris, France. Occurs in caliche ($NaNO_3$) (the iodine is present as $NaIO_3$). The iodine is precipitated by acidified iodate after reduction by sodium hydrogensulphite solution. The iodine is purified by sublimation. Iodine can also be found in sea weeds.

Properties

Iodine is a dark black shiny solid (I_2) which is very volatile, producing a violet vapour on heating (sublimes easily). The vapour has an unpleasant smell and it is poisonous. Iodine is a relatively unreactive halogen but will produce iodides with metals such as iron. It dissolves readily in organic solvents.

Uses	Iodine is used in the manufacture of medicines (KI), disinfectants (CHI_3), photographic chemicals (AgI), organic chemicals (C_6H_5I), food supplements and quartz–halogen lamps. ^{125}I is used as a medical tracer.

Biological Role	Iodine is an essential trace element. The daily requirement being 0.0003 mg. It is essential for the production of the hormone thyroxine, a vital metabolic regulator. Deficiency of iodine causes goitre and cretinism both of which can be overcome by adding iodine to the diet.

Melting Point/K	386.7	Boiling Point/K	457.5	Density/g cm^{-3}	4.93

Electronic Configuration	[Kr] $4d^{10}$, $5s^2$ $5p^5$

Oxidation State(s)	$\underline{-1}$,0,+1,+3,+5,$\underline{+7}$	Isotopes	^{127}I(100%)

S.E.P.	$E^{\ominus}$/Volts
IO_3^- | I_2	+ 1.20 Acid solution

Electronegativity (Pauling)	2.5

Radii	nm
Ionic	0.216 (−1)
Atomic	0.133
Van der Waals	0.215
Covalent	0.133

Enthalpy Data	kJ mol^{-1}
1st Ionisation Energy	1008
2nd Ionisation Energy	1846
1st Electron Affinity	−296.2
2nd Electron Affinity	–
Enthalpy of Fusion, $\Delta H^{\ominus}_{Fusion}$	7.9
Enthalpy of Vaporisation, $\Delta H^{\ominus}_{Vaporisation}$	20.9
Enthalpy of Atomisation, $\Delta H^{\ominus}_{Atomisation}$	106.8
	J K^{-1} mol^{-1}
Standard Entropy, $S^{\ominus}$	116.2

Associated Bond Lengths (nm) & Energies (kJ mol^{-1})	I–I(0.267,151), I–H(0.161,299), I–O(0.195,234), I–Cl(0.232,208), I– C(0.213,218)

Iridium

[Latin, 'iris' = rainbow]

Atomic Number	77
Relative Atomic Mass	192.22

Chemical Symbol	Ir	Group	–	Period	6 (3rd series T.M.)

Main Source & History

Discovered by S. Tennant in 1803, London, UK. It occurs in osmiridium a native platinum alloy (52.5% Ir). It is separated by electrolysis followed by ion exchange and solvent extraction to give pure iridium.

Properties

Iridium is an extremely hard, silvery metal which is lustrous in appearance. It resembles platinum (occuring with it). Iridium is resistant to the chemical action of air, water and all acids.

Uses

Iridium is used mainly in the manufacture of alloys which are used for fountain pen nibs (with Pt and Os), thermocouples (Pt/Ir and Ir/Rh) and spark plugs. Alloys containing 10% iridium and 90% platinum are used as standard measures of mass and length. Iridium is also used as a catalyst, for example in the manufacture of nitric acid (Pt/Ir).

Biological Role

Iridium has no known biological role.

Melting Point/K	2683.0	Boiling Point/K	4403.0	Density/g cm^{-3}	22.42

Electronic Configuration	[Xe] $4f^{14}$, $5d^{7}$, $6s^{2}$

Oxidation State(s)	– 1, 0, +1, +2, <u>+3</u>, <u>+4</u>, +5, +6	Isotopes	^{193}Ir(61.5%), ^{191}Ir(38.5%)

S.E.P.	$E^{\ominus}$/Volts
Ir^{3+}|Ir	+ 1.16

Electronegativity (Pauling)	2.2

Radii	nm
Ionic	0.075 (+3)
Atomic	0.136
Van der Waals	–
Covalent	0.126

Enthalpy Data	kJ mol^{-1}
1st Ionisation Energy	880
2nd Ionisation Energy	1550
1st Electron Affinity	+160.0
2nd Electron Affinity	–
Enthalpy of Fusion, $\Delta H^{\ominus}_{Fusion}$	26.4
Enthalpy of Vaporisation, $\Delta H^{\ominus}_{Vaporisation}$	612.1[2]
Enthalpy of Atomisation, $\Delta H^{\ominus}_{Atomisation}$	665.2
	J $K^{-1}mol^{-1}$
Standard Entropy, $S^{\ominus}$	35.5

Associated Bond Lengths (nm) & Energies (kJ mol^{-1})	n.a.

Iron

[Anglo – Saxon, iron]

Atomic Number	26
Relative Atomic Mass	55.847

Chemical Symbol	Fe	Group	–	Period	4 (1st series T.M.)

Main Source & History

Iron has been known since pre – historic times. The main ores are haematite (Fe_2O_3) and magnetite (Fe_3O_4). Iron is the most important of all metals and is extracted from the oxide by reduction with carbon in the blast furnace. Limestone being added to remove acid impurities.

Properties

Pure iron is a silvery – white, magnetic metal which is soft and lustrous in appearance. Iron is a moderately reactive metal which corrodes (rusts) in damp air and reacts with dilute acids liberating hydrogen. It is rendered passive by concentrated nitric acid.

Uses

The vast majority of iron is converted into steels. These alloys are used in the construction industry; to make stainless steels, for cutlery and surgical instruments (Fe/Cr/C/Ni); tungsten steel, for high speed cutting tools (Fe/W/C); invar for making watches (Fe/Ni). It is also used in the production of electromagnets. Compounds of iron are used in medicines ($FeSO_4$), as pigments in paints and dyes (Fe_2O_3).

Biological Role

Iron is an essential element. The daily requirement is 0.01 mg. It is needed for the production of haemoglobin, the red substance of blood which carries oxygen to the body tissues. Deficiency of iron causes anaemia, which can be treated by addition of iron to the diet. In plants iron is required for the synthesis of chlorophyll.

Melting Point/K	1808.0	Boiling Point/K	3023.0	Density/g cm^{-3}	7.87

Electronic Configuration	[Ar] $3d^6$, $4s^2$

Oxidation State(s)	–2, –1, 0, +1, +2, +3, +4, +5, +6	Isotopes	^{56}Fe(91.7%), ^{54}Fe(5.8%)

S.E.P.	$E^{\ominus}$/Volts
$Fe^{3+}\|Fe$	– 0.04 (Acid solution)

Electronegativity (Pauling)	1.8

Radii	nm
Ionic	0.065 (+3)
Atomic	0.124
Van der Waals	–
Covalent	0.116

Enthalpy Data	kJ mol^{-1}
1st Ionisation Energy	759
2nd Ionisation Energy	1561
1st Electron Affinity	+24.0
2nd Electron Affinity	–
Enthalpy of Fusion, $\Delta H^{\ominus}_{Fusion}$	15.4
Enthalpy of Vaporisation, $\Delta H^{\ominus}_{Vaporisation}$	351.0
Enthalpy of Atomisation, $\Delta H^{\ominus}_{Atomisation}$	416.3
	J $K^{-1}mol^{-1}$
Standard Entropy, $S^{\ominus}$	27.3

Associated Bond Lengths (nm) & Energies (kJ mol^{-1})	n.a.

Krypton

[Greek, 'kryptos' = hidden]

Atomic Number	36
Relative Atomic Mass	83.80

Chemical Symbol	Kr	Group	0 (8)	Period	4

Main Source & History

Discovered by Sir W. Ramsay and M.W. Travers in 1890, London, UK. It can be obtained in its pure form by the fractional distillation of liquid air.

Properties

It is a colourless, odourless, relatively inert gas. It is a monatomic gas, and there are no interactions between the atoms apart from weak van der Waals forces. KrF_2 and KrF_4 have been synthesised.

Uses	Krypton is used in electrical discharge tubes, with argon. It is also used as a lasing element, and is present in some types of photographic flash units.

Biological Role	Krypton has no known biological role. It is known be to non–toxic.

Melting Point/K	116.6	Boiling Point/K	121.0	Density/g cm^{-3}	2.15^{121K}

Electronic Configuration	[Ar] $3d^{10}$, $4s^2$ $4p^6$

Oxidation State(s)	0, +2, +4	Isotopes	^{84}Kr(57.0%), ^{86}Kr(17.3%)

S.E.P. $E^{\ominus}$/Volts
n.a.

Electronegativity (Pauling)	–

Radii	nm
Ionic	–
Atomic	0.110
Van der Waals	0.200
Covalent	0.189

Enthalpy Data	kJ mol^{-1}
1st Ionisation Energy	1351
2nd Ionisation Energy	2368
1st Electron Affinity	0.0
2nd Electron Affinity	–
Enthalpy of Fusion, $\Delta H^{\ominus}_{Fusion}$	1.6
Enthalpy of Vaporisation, $\Delta H^{\ominus}_{Vaporisation}$	9.0
Enthalpy of Atomisation, $\Delta H^{\ominus}_{Atomisation}$	0.0
	J $K^{-1}mol^{-1}$
Standard Entropy, $S^{\ominus}$	164.0

Associated Bond Lengths (nm) & Energies (kJ mol^{-1})	Kr–F(0.189,50)

Lanthanum

[Greek, 'lanthano' = conceal]

Atomic Number	57
Relative Atomic Mass	138.9055

Chemical Symbol	La	Group	–	Period	6 Lanthanide series

Main Source & History

Discovered by C.G. Mosander in 1839, Stockholm, Sweden. The main ores are monazite and bastnaesite from which lanthanum can be extracted with difficulty by ion exchange. The metal may be obtained by reduction of $LaCl_3$ with calcium.

Properties

Lanthanum is a soft, sivery–white metal which is a good conductor of heat and electricity. Lanthanum reacts with water liberating hydrogen. It reacts rapidly with air as well as burning easily when ignited. Lanthanum also reacts with acids.

Uses

Lanthanum metal is used in the manufacture of alloys with aluminium and magnesium, as well as in steels. Compounds of lanthanum are used in the production of optical glasses, as high temperature refractory linings and 'flints'.

Biological Role

Lanthanum has no known biological role although the element is non–toxic in small amounts. The La^{3+} ion has been used of late as a biological tracer for calcium.

Melting Point/K	1194.0	Boiling Point/K	3730.0	Density/g cm^{-3}	6.14

Electronic Configuration	[Xe] $5d^1$, $6s^2$

Oxidation State(s)	0,±3	Isotopes	^{139}La(99.91%), ^{138}La(0.09%)

S.E.P.	$E^\ominus$/Volts
La^{3+}|La	– 2.38 (Acid solution)

Electronegativity (Pauling)	1.1

Radii	nm
Ionic	0.116 (+3)
Atomic	0.188
Van der Waals	–
Covalent	0.169

Enthalpy Data	kJ mol^{-1}
1st Ionisation Energy	538
2nd Ionisation Energy	1067
1st Electron Affinity	+53.0
2nd Electron Affinity	–
Enthalpy of Fusion, $\Delta H^\ominus_{Fusion}$	11.0
Enthalpy of Vaporisation, $\Delta H^\ominus_{Vaporisation}$	400.0
Enthalpy of Atomisation, $\Delta H^\ominus_{Atomisation}$	427.0
	J $K^{-1}mol^{-1}$
Standard Entropy, $S^\ominus$	56.9

Associated Bond Lengths (nm) & Energies (kJ mol^{-1})	n.a.

Lawrencium

[After E.O. Lawrence]

Atomic Number	103
Relative Atomic Mass	260.1053[1]

Chemical Symbol	Lr	Group	–	Period	7 Actinide series

Main Source & History

Discovered by A. Ghiorso, A.E. Larch, R.M. Latimer and T. Sikkeland in 1961, California, USA. It can be formed by bombarding ^{252}Cf with ^{10}B and ^{11}B nuclei.

Properties

This transuranic element is a radioactive metal. Only a few atoms of lawrencium have ever been produced. The most stable isotope is ^{260}Lr which has a $T_{1/2}$ of 3 minutes.

Uses	Lawrencium has no known uses.

Biological Role	Lawrencium has no biological role. It is hazardous to health due to its extreme radioactive nature.

Melting Point/K	–	Boiling Point/K	–	Density/g cm^{-3}	–

Electronic Configuration	[Rn] $5f^{14}$, $6d^1$, $7s^2$

Oxidation State(s)	0,$\underline{\pm 3}$	Isotopes	^{260}Lr(0%), [2 others]

S.E.P.	$E^{\ominus}$/Volts
Lr^{3+}|Lr	– 2.06

Electronegativity (Pauling)	1.30

Radii	nm
Ionic	0.094 (+3)
Atomic	–
Van der Waals	–
Covalent	–

Enthalpy Data	kJ mol^{-1}
1st Ionisation Energy	–
2nd Ionisation Energy	–
1st Electron Affinity	–
2nd Electron Affinity	–
Enthalpy of Fusion, $\Delta H^{\ominus}_{Fusion}$	–
Enthalpy of Vaporisation, $\Delta H^{\ominus}_{Vaporisation}$	–
Enthalpy of Atomisation, $\Delta H^{\ominus}_{Atomisation}$	–
	J $K^{-1}mol^{-1}$
Standard Entropy, $S^{\ominus}$	–

Associated Bond Lengths (nm) & Energies (kJ mol^{-1})	n.a.

Lead

[Anglo–Saxon, lead]

Atomic Number	82
Relative Atomic Mass	207.2

Chemical Symbol	Pb	Group	4	Period	6

Main Source & History

Lead has been known since pre–historic times. Its main ores are galena (PbS) and cerussite ($PbCO_3$). These ores are roasted to give PbO which is then reduced to lead with carbon.

Properties

Lead is a dull grey–blue metal. It is very malleable and ductile. All natural decay series are completed at lead. Lead is relatively unreactive towards water and oxygen separately but corrodes in moist air. Lead is attacked by concentrated acids and alkalis.

Uses

Lead is used in the production of alloys, eg. solder and pewter, as radiation shielding and in storage batteries. The compounds are used as petrol additives (tetraethyl lead to a decreasing extent), pigments in paints ($PbCl_2$), in the production of high quality glass (PbO), insecticides ($Pb_3(AsO_4)_2$)and as mordants ($Pb(NO_3)_2$).

Biological Role

Lead has no known biological role. The element and its compounds are highly toxic. There is a statuatory limit of 2ppm for lead in food. The presence of lead can cause malformation of the foetus.

Melting Point/K	601.0	Boiling Point/K	2013.0	Density/g cm^{-3}	11.34

Electronic Configuration	[Xe] $4f^{14}$, $5d^{10}$, $6s^2$ $6p^6$

Oxidation State(s)	– 2,0,<u>+2</u>,+4	Isotopes	^{208}Pb(52.3%), ^{206}Pb(25.1%)

S.E.P.	$E^{\ominus}$/Volts
Pb^{2+}|Pb	– 0.13 (Acid solution)

Electronegativity (Pauling)	1.8

Radii	nm
Ionic	0.119 (+2)
Atomic	0.175
Van der Waals	0.200
Covalent	0.154

Enthalpy Data	kJ mol^{-1}
1st Ionisation Energy	716
2nd Ionisation Energy	1450
1st Electron Affinity	–35.2
2nd Electron Affinity	–
Enthalpy of Fusion, $\Delta H^{\ominus}_{Fusion}$	5.1
Enthalpy of Vaporisation, $\Delta H^{\ominus}_{Vaporisation}$	178.0
Enthalpy of Atomisation, $\Delta H^{\ominus}_{Atomisation}$	196.0
	J $K^{-1}mol^{-1}$
Standard Entropy, $S^{\ominus}$	64.8

Associated Bond Lengths (nm) & Energies (kJ mol^{-1})	Pb–Pb(0.350,100), Pb–H(0.184,180), Pb–O(0.192,398), Pb–Cl(0.247,244), Pb–C(0.229,130)

Lithium

[Greek, 'lithos' = stone]

Atomic Number	3
Relative Atomic Mass	6.941

Chemical Symbol	Li	Group	1	Period	2

Main Source & History

Discovered by J.A. Arfvedson in 1817, Stockholm, Sweden. Important minerals are spodumene ($LiAlSi_2O_6$) and petalite ($LiAlSi_4O_{10}$). The element is obtained by electrolytic reduction of the fused chloride.

Properties

Lithium is a soft, silvery coloured metal which is a good conductor of heat and electricity. It is the lightest solid element. Lithium is a reactive element stored under oil due to its rapid reaction with oxygen and water (liberating hydrogen).

Uses

Lithium is used in the manufacture of low melting point alloys and in nuclear power stations as a heat transfer agent. Compounds of lithium are used in medicine (Li_2CO_3 anti−depressant), in glass and ceramic manufacture (Li_2CO_3), air conditioning units in submarines and spacecraft to absorb carbon dioxide (LiOH), in fungicides and in batteries.

Biological Role

Lithium has no known biological role. It has relatively low toxicity but it can cause malformation of the foetus.

Melting Point/K	454.0	Boiling Point/K	1620.0	Density/g cm^{-3}	0.53

Electronic Configuration	[He] $2s^1$

Oxidation State(s)	−1,0,$\underline{+1}$	Isotopes	^7Li(92.6%), ^6Li(7.4%)

S.E.P.	$E^{\ominus}$/Volts
Li^+|Li	− 3.04

Electronegativity (Pauling)	1.0

Radii	nm
Ionic	0.060 (+1)
Atomic	0.152
Van der Waals	0.180
Covalent	0.123

Enthalpy Data	kJ mol^{-1}
1st Ionisation Energy	520
2nd Ionisation Energy	7298
1st Electron Affinity	−59.8
2nd Electron Affinity	–
Enthalpy of Fusion, $\Delta H^{\ominus}_{Fusion}$	3.0
Enthalpy of Vaporisation, $\Delta H^{\ominus}_{Vaporisation}$	134.7
Enthalpy of Atomisation, $\Delta H^{\ominus}_{Atomisation}$	159.4
	J $K^{-1} mol^{-1}$
Standard Entropy, $S^{\ominus}$	29.1

Associated Bond Lengths (nm) & Energies (kJ mol^{-1})	n.a.

Lutetium

[Latin, 'Lutetia' = Paris]

Atomic Number	71
Relative Atomic Mass	174.967

Chemical Symbol	Lu	Group	–	Period	6 Lanthanide series

Main Source & History

Discovered by G. Urbain (France) and C. James (USA) in 1907. The main ores are monazite and bastnaesite from which lutetium is extracted with extreme difficulty by ion exchange.

Properties

Lutetium, a silvery white metal, is one of the rarest of the lanthanides. It is also the hardest and most dense and is a good conductor of heat and electricity. Lutetium is slowly attacked by air and water (liberating hydrogen) and is also attacked by acids.

Uses

Lutetium nuclides are used in catalytic processes, cracking and polymerisation. After neutron irradiation they are a source of pure beta radiation.

Biological Role

Lutetium has no known biological role.

Melting Point/K	1929.0	Boiling Point/K	3668.0	Density/g cm^{-3}	9.84

Electronic Configuration	[Xe] $4f^{14}$, $5d^{1}$, $6s^{2}$

Oxidation State(s)	0, ± 3	Isotopes	^{175}Lu(97.4%), ^{176}Lu(2.6%)

S.E.P.	$E^{\ominus}$/Volts
Lu^{3+}|Lu	– 2.30 (Acid solution)

Electronegativity (Pauling)	1.2

Radii	nm
Ionic	0.085 (+3)
Atomic	0.172
Van der Waals	–
Covalent	0.156

Enthalpy Data	kJ mol^{-1}
1st Ionisation Energy	524
2nd Ionisation Energy	1340
1st Electron Affinity	+50.0[2]
2nd Electron Affinity	–
Enthalpy of Fusion, $\Delta H^{\ominus}_{Fusion}$	19.2
Enthalpy of Vaporisation, $\Delta H^{\ominus}_{Vaporisation}$	428.0
Enthalpy of Atomisation, $\Delta H^{\ominus}_{Atomisation}$	–
	J K^{-1} mol^{-1}
Standard Entropy, $S^{\ominus}$	51.0

Associated Bond Lengths (nm) & Energies (kJ mol^{-1})	n.a.

Magnesium

[Greek, 'Magnesia' = a region in Thessaly]

Atomic Number	12
Relative Atomic Mass	24.3050

Chemical Symbol	Mg	Group	2	Period	3

Main Source & History

First isolated by Sir H. Davy in 1808, London, UK. The important ores are dolomite ($MgCaC_2O_6$), magnesite ($MgCO_3$) and carnalite ($KMgCl_3.6H_2O$). The metal is obtained by electrolysis of the fused halide or by thermal reduction of the oxide.

Properties

Magnesium is a silvery–white, relatively soft metal. It is lustorous in appearance and is a good conductor of heat and electricity. Magnesium is a reactive metal which tarnishes rapidly in air and burns, when ignited, giving a brilliant white flame. It also reacts with steam and acids liberating hydrogen.

Uses	Magnesium has extensive uses including manufacture of many light alloys (duralumin for aircraft construction), in batteries, in the extraction of metals (eg. titanium), and as a sacrificial anode. Compounds of magnesium are used in glass and ceramics manufacture ($MgCO_3$), medicines ('Milk of Magnesia', $Mg(OH)_2$, Epsom Salts, $MgSO_4.7H_2O$), in the paper (MgO) and sugar ($Mg(OH)_2$) industries and in water treatment ($MgOH)_2$).
Biological Role	Magnesium is an essential element. The daily requirement is 0.2 – 0.3 mg. Along with calcium and phosphorus, magnesium is used to form the bone structure. It is also necessary for muscle contraction to take place. Also it is necessary for certain hormones, eg. insulin (which controls sugar metabolism). Deficiency of magnesium leads to fits. In plants, magnesium is a constituent of chlorophyll.

Melting Point/K	922.0	Boiling Point/K	1380.0	Density/g cm^{-3}	1.74

Electronic Configuration	[Ne] $3s^2$

Oxidation State(s)	0, +1, $\underline{+2}$	Isotopes	^{24}Mg(78.6%), ^{26}Mg(11.01%)

S.E.P.	$E^\ominus$/Volts
Mg^{2+}|Mg	– 2.36 (Acid solution)

Electronegativity (Pauling)	1.2

Radii	nm
Ionic	0.065 (+2)
Atomic	0.160
Van der Waals	0.170
Covalent	0.136

Enthalpy Data	kJ mol^{-1}
1st Ionisation Energy	738
2nd Ionisation Energy	1451
1st Electron Affinity	+67.0
2nd Electron Affinity	–
Enthalpy of Fusion, $\Delta H^\ominus_{Fusion}$	9.0
Enthalpy of Vaporisation, $\Delta H^\ominus_{Vaporisation}$	128.7
Enthalpy of Atomisation, $\Delta H^\ominus_{Atomisation}$	147.7
	J $K^{-1}mol^{-1}$
Standard Entropy, $S^\ominus$	32.7

Associated Bond Lengths (nm) & Energies (kJ mol^{-1})	n.a.

Manganese

[Latin, 'magnes' = magnet]

Atomic Number	25
Relative Atomic Mass	54.9381

Chemical Symbol	Mn	Group	–	Period	4 1st series T.M.

Main Source & History

First isolated by J.G. Gahn in1774, Stockholm, Sweden, a purer sample was isolated by J.F. John in 1807. The main ores are pyrolusite (MnO_2) and hausmannite (Mn_3O_4) from which manganese is extracted by reduction with aluminium or carbon. It is purified by elctrolysis.

Properties

Manganese, when pure, is a silvery–white metal. It is brittle and intensely hard, being equal to hardened and tempered tool steel. It is a fairly reactive metal combining with a variety of non–metals on heating. It liberates hydrogen from water and dilute acids.

Uses

Manganese is used extensively in the production of manganese steels with iron, such as those used in making rock drills and railway points. All steels contain a little manganese. Compounds of manganese are used as oxidising agents ($KMnO_4$), in batteries (MnO_2), in paints and dyes (MnO_2), fertilisers, fungicides and herbicides.

Biological Role

Manganese is an essential trace element. Manganese activates the enzymes alkaline phosphatase and organase which are concerned with bone and urea formation respectively. Deficiency of manganese can cause deformaties of the skeleton and sterility, In plants it activates carboxylases.

Melting Point/K	1517.0	Boiling Point/K	2235.0	Density/g cm^{-3}	7.2

Electronic Configuration	[Ar] $3d^5$, $4s^2$

Oxidation State(s)	–1,0,+1,$\underline{+2}$,+3,+4,+5,+6,$\underline{+7}$	Isotopes	^{55}Mn(100%)

S.E.P.	$E^\ominus$/Volts
Mn^{2+}|Mn	– 1.18 Acid solution

Electronegativity (Pauling)	1.5

Radii	nm
Ionic	0.080 (+2)
Atomic	0.137
Van der Waals	–
Covalent	0.117

Enthalpy Data	kJ mol^{-1}
1st Ionisation Energy	717
2nd Ionisation Energy	1509
1st Electron Affinity	+94.0
2nd Electron Affinity	–
Enthalpy of Fusion, $\Delta H^\ominus_{Fusion}$	14.6
Enthalpy of Vaporisation, $\Delta H^\ominus_{Vaporisation}$	220.0
Enthalpy of Atomisation, $\Delta H^\ominus_{Atomisation}$	280.7
	J $K^{-1}mol^{-1}$
Standard Entropy, $S^\ominus$	32.0

Associated Bond Lengths (nm) & Energies (kJ mol^{-1})	n.a.

Mendelevium

[After D.I. Mendeleev]

Atomic Number	101
Relative Atomic Mass	258.0986[1]

Chemical Symbol	Md	Group	–	Period	7 Actinide series

Main Source & History

Discovered by G.R. Choppin, A. Ghiorso, B.G. Harvey, G.T. Seaborg and S.G. Thompson in 1955, California, USA. Can be formed by bombarding ^{253}Es with alpha particles.

Properties

This transuranic element is a radioactive metal. Only a few atoms of mendelevium have ever been produced. The most stable isotope ^{258}Md has a $T_{1/2}$ of 54 days.

Uses	Mendelevium has no known uses.

Biological Role	Mendelevium has no known biological role. It is hazardous to health due to its radioactive nature.

Melting Point/K	–	Boiling Point/K	–	Density/g cm^{-3}	–

Electronic Configuration	[Rn] $5f^{13}$, $7s^2$

Oxidation State(s)	0,$\underline{+3}$	Isotopes	–

S.E.P.	$E^{\ominus}$/Volts
Md^{3+}|Md	– 1.66 Acid solution

Electronegativity (Pauling)	1.3

Radii	nm
Ionic	0.096 (+3)
Atomic	–
Van der Waals	–
Covalent	–

Enthalpy Data	kJ mol^{-1}
1st Ionisation Energy	635
2nd Ionisation Energy	–
1st Electron Affinity	–
2nd Electron Affinity	–
Enthalpy of Fusion, $\Delta H^{\ominus}_{Fusion}$	–
Enthalpy of Vaporisation, $\Delta H^{\ominus}_{Vaporisation}$	–
Enthalpy of Atomisation, $\Delta H^{\ominus}_{Atomisation}$	–
	J $K^{-1}mol^{-1}$
Standard Entropy, $S^{\ominus}$	–

Associated Bond Lengths (nm) & Energies (kJ mol^{-1})	n.a.

Mercury

[After Greek god Mercurius]

Atomic Number	80
Relative Atomic Mass	200.59

Chemical Symbol	Hg	Group	–	Period	6 3rd series T.M.

Main Source & History

Mercury has been known since 300 BC. The most important mercury ore is cinnabar (HgS) from which mercury is obtained by roasting in air. The mercury distils at the roasting temperature and is then condensed. It is purified under reduced pressure by distillation.

Properties

Mercury is the only liquid metal at room temperature. It is a dense, silvery coloured element. Pure mercury is not attacked by air or water at ordinary temperatures. It is unreactive to acids unless they are oxidising.

Uses

Mercury is used in electrical switches, thermometers, barometers, arc lights and in the chloro–alkali industry where it is used as the cathode in the production of sodium hydroxide and chlorine during the electrolysis of brine. Sodium and zinc amalgams (alloys with mercury) are used as reducing agents and tin amalgam is used as a dental filling. Compounds of mercury are used to make fungicides, insecticides (Hg_2Cl_2), timber preservatives and detonators ($Hg(ONC)_2$).

Biological Role

Mercury has no known biological role. It is known to accumulate in the body forming dangerously toxic organo–mercury compounds. It can cause malformation of the foetus.

Melting Point/K	234.0	Boiling Point/K	630.0	Density/g cm^{-3}	13.55

Electronic Configuration	[Xe] $4f^{14}$, $5d^{10}$, $6s^2$

Oxidation State(s)	0,±1,±2	Isotopes	^{202}Hg(29.8%), ^{200}Hg(23.1%)

S.E.P.	$E^{\ominus}$/Volts
Hg^{2+}|Hg	+ 0.85 Acid solution

Electronegativity (Pauling)	2.0

Radii	nm
Ionic	0.112 (+2)
Atomic	0.150
Van der Waals	–
Covalent	0.144

Enthalpy Data	kJ mol^{-1}
1st Ionisation Energy	1007
2nd Ionisation Energy	1810
1st Electron Affinity	+148.6
2nd Electron Affinity	–
Enthalpy of Fusion, $\Delta H^{\ominus}_{Fusion}$	2.3
Enthalpy of Vaporisation, $\Delta H^{\ominus}_{Vaporisation}$	59.2
Enthalpy of Atomisation, $\Delta H^{\ominus}_{Atomisation}$	61.3
	J $K^{-1}mol^{-1}$
Standard Entropy, $S^{\ominus}$	76.0

Associated Bond Lengths (nm) & Energies (kJ mol^{-1})	n.a.

Molybdenum

[Greek, 'molybdos' = lead]

Atomic Number	42
Relative Atomic Mass	95.94

Chemical Symbol	Mo	Group	–	Period	5 2nd series T.M.

Main Source & History

First isolated by P.J. Hjelm in 1790, Uppsala, Sweden. The main ore is molybdenite (MoS_2) which is roasted to the oxide and reduced to the metal with hydrogen. It is also formed as a by–product in the production of copper.

Properties

Molybdenum is a fairly soft, lustrous, silvery–white metal. The element is only slowly attacked by acids whilst remaining unchanged when exposed to oxygen and water at ordinary temperatures.

Uses

Molybdenum is used particularly in the formation of alloy steels, eg. Ni/Mo steels for making propellor shafts and Cr/Mo steel to a small extent in aircraft production. The most important use is in the production of alloys to make high speed cutting tools. Compounds of molybdenum are used as pigments in paints (MoO_3) and as solid lubricants (MoS_2).

Biological Role

Molybdenum is an essential trace element, the daily requirement being extremely small. Molybdenum is a constituent of xanthine oxidase, the enzyme which plays a part in the metabolism of purines which make up two of the bases of the DNA code. It can cause malformation of the foetus in large quantities. In plants it activates certain enzymes in the nitrogen metabolism.

Melting Point/K	2883.0	Boiling Point/K	4885.0	Density/g cm^{-3}	10.28

Electronic Configuration	[Kr] $4d^5$, $5s^1$

Oxidation State(s)	0, +2, +3, +4, +5, $\underline{+6}$	Isotopes	^{98}Mo(24.1%), ^{96}Mo(16.7%)

S.E.P.	$E^\ominus$/Volts
H_2MoO_4|Mo	+ 0.11 Acid solution

Electronegativity (Pauling)	1.8

Radii	nm
Ionic	0.061 (+6)
Atomic	0.136
Van der Waals	–
Covalent	0.130

Enthalpy Data	kJ mol^{-1}
1st Ionisation Energy	694
2nd Ionisation Energy	1558
1st Electron Affinity	+100.0
2nd Electron Affinity	–
Enthalpy of Fusion, $\Delta H^\ominus_{Fusion}$	27.6
Enthalpy of Vaporisation, $\Delta H^\ominus_{Vaporisation}$	590.0
Enthalpy of Atomisation, $\Delta H^\ominus_{Atomisation}$	658.1
	J $K^{-1}mol^{-1}$
Standard Entropy, $S^\ominus$	28.7

Associated Bond Lengths (nm) & Energies (kJ mol^{-1})	n.a.

Neodymium

[Greek, 'neos didymos' = new twin]

Atomic Number	60
Relative Atomic Mass	144.24

Chemical Symbol	Nd	Group	–	Period	6 Lanthanide series

Main Source & History

Discovered by C. Auer von Welsbach in 1885, Vienna, Austria. The main ores are monazite and bastnaesite from which neodymium is obtained, with difficulty, by ion exchange. The metal may be obtained by reduction of $NdCl_3$ with calcium.

Properties

Neodymium is a silvery–white metal which is a good conductor of heat and electricity. The element reacts with air and liberates hydrogen slowly from water. It also reacts with acids.

Uses	Neodymium is used in the manufacture of glass, glazes and electronic components, eg. capacitors. It is also used in the production of alloys and 'flints'.

Biological Role	Neodymium has no known biological role.

Melting Point/K	1294.0	Boiling Point/K	3341.0	Density/g cm^{-3}	7.00

Electronic Configuration	[Xe] $4f^4$, $6s^2$

Oxidation State(s)	0, +2, $\underline{+3}$, +4	Isotopes	^{142}Nd(27.2%), ^{144}Nd(23.8%)

S.E.P.	$E^{\ominus}$/Volts
Nd^{3+}|Nd	– 2.32 Acid solution

Electronegativity (Pauling)	1.2

Radii	nm
Ionic	0.104 (+3)
Atomic	0.182
Van der Waals	–
Covalent	0.164

Enthalpy Data	kJ mol^{-1}
1st Ionisation Energy	530
2nd Ionisation Energy	1035
1st Electron Affinity	+50.0[2]
2nd Electron Affinity	–
Enthalpy of Fusion, $\Delta H^{\ominus}_{Fusion}$	7.1
Enthalpy of Vaporisation, $\Delta H^{\ominus}_{Vaporisation}$	328.0
Enthalpy of Atomisation, $\Delta H^{\ominus}_{Atomisation}$	–
	J $K^{-1}mol^{-1}$
Standard Entropy, $S^{\ominus}$	71.5

Associated Bond Lengths (nm) & Energies (kJ mol^{-1})	n.a.

Neon

[Greek, 'neos' = new]

Atomic Number	10
Relative Atomic Mass	20.1797

Chemical Symbol	Ne	Group	0 (8)	Period	2

Main Source & History

Discovered by Sir W. Ramsay and M.W. Travers in 1898, London, UK. Obtained by the fractional distillation of liquid air.

Properties

Neon is a colourless, tasteless, odourless, inert gas. It is a monatomic gas, and there are no interactions between the atoms apart from weak van der Waals forces.

Uses

Neon is used extensively in discharge tubes on its own or mixed with a little mercury vapour, in Geiger–Muller tubes and in helium–neon gas lasers.

Biological Role

Neon has no known biological role. It is known be to non–toxic.

Melting Point/K	24.5	Boiling Point/K	27.1	Density/g cm^{-3}	1.20^{27K}

Electronic Configuration	[He] $2s^2$ $2p^6$

Oxidation State(s)	0	Isotopes	^{20}Ne(90.5%), ^{22}Ne(9.2%)

S.E.P. $E^\ominus$/Volts
n.a.

Electronegativity (Pauling)	–

Radii	nm
Ionic	–
Atomic	0.065
Van der Waals	0.160
Covalent	–

Enthalpy Data	kJ mol^{-1}
1st Ionisation Energy	2081
2nd Ionisation Energy	3952
1st Electron Affinity	0.0
2nd Electron Affinity	–
Enthalpy of Fusion, $\Delta H^\ominus_{Fusion}$	0.33
Enthalpy of Vaporisation, $\Delta H^\ominus_{Vaporisation}$	1.8
Enthalpy of Atomisation, $\Delta H^\ominus_{Atomisation}$	0.0
	J $K^{-1}mol^{-1}$
Standard Entropy, $S^\ominus$	146.2

Associated Bond Lengths (nm) & Energies (kJ mol^{-1})	n.a.

Neptunium

[After the planet Neptune]

Atomic Number	93
Relative Atomic Mass	237.0482[1]

Chemical Symbol	Np	Group	–	Period	7 Actinide series

Main Source & History

Discovered by P. Abelson and E.M. McMillan in 1940, California, USA. Trace amounts occur naturally being formed by neutron capture by ^{238}U (naturally occuring). Metallic ^{237}Np is obtained as a by–product by nuclear reactors in the production of plutonium. Neptunium can be separated by solvent extraction and obtained by reduction of NpF_3 with lithium.

Properties

This transuranic element is a radioactive, silvery–white metal. The most stable isotope ^{237}Np has a $T_{1/2}$ of 2.14 x 10^6 years. Neptunium reacts with oxygen as well as liberating hydrogen from steam. It also reacts with acids.

Uses	Neptunium has no known uses.
Biological Role	Neptunium has no known biological role. It is hazardous to health due to its radioactive nature.

Melting Point/K	913.0	Boiling Point/K	4175.0	Density/g cm^{-3}	20.47 ^{293K}

Electronic Configuration	[Rn] $5f^4,6d^1\ 7s^2$

Oxidation State(s)	0, +3, +4, <u>+5</u>, +6, +7	Isotopes	$^{237}Np(0\%)$, $^{239}Np(0\%)$

S.E.P.	$E^{\ominus}$/Volts
Np^{3+}|Np	– 1.79 (Acid solution)

Electronegativity (Pauling)	1.4

Radii	nm
Ionic	0.088 (+5)
Atomic	0.131
Van der Waals	–
Covalent	–

Enthalpy Data	kJ mol^{-1}
1st Ionisation Energy	597
2nd Ionisation Energy	–
1st Electron Affinity	–
2nd Electron Affinity	–
Enthalpy of Fusion, $\Delta H^{\ominus}_{Fusion}$	9.5
Enthalpy of Vaporisation, $\Delta H^{\ominus}_{Vaporisation}$	336.6
Enthalpy of Atomisation, $\Delta H^{\ominus}_{Atomisation}$	–
	J $K^{-1}mol^{-1}$
Standard Entropy, $S^{\ominus}$	–

Associated Bond Lengths (nm) & Energies (kJ mol^{-1})	n.a.

Nickel

[German, 'kupfernickel' = devil's copper]

Atomic Number	28
Relative Atomic Mass	58.69

Chemical Symbol	Ni	Group	–	Period	4 1st series T.M.

Main Source & History

Discovered by A.F. Cronstedt in 1751, Stockholm, Sweden. The main ores are millerite (NiS) and pentlandite ($Fe_9Ni_9S_8$). The ores are roasted to give nickel(II) oxide which is then reduced to the metal with carbon or hydrogen. The metal is purified electrolytically or by the Mond Process.

Properties

Nickel is a moderately hard, silvery–white metal. It is lustrous in appearance and both malleable and ductile. The metal is resistant to corrosion. It is not affected by water. It dissolves in acids but not concentrated nitric acid.

Uses

Nickel is used as a catalyst in the hydrogenation of alkenes. It is also used in the production of alloys such as stainless steel, alnico (to make permanent magnets), nichrome (windings in electric motors) and Monel metal used to make propeller shafts. Cupro–nickel is used to make coins. Compounds of nickel are used in electroplating ($Ni(NH_4)_2(SO_4)_2.6H_2O$) and as pigments (NiO).

Biological Role

The biological role of nickel is uncertain even though it is found in human tissue. Some nickel compounds are thought to be carcinogenic.

Melting Point/K	1726.0	Boiling Point/K	3005.0	Density/g cm^{-3}	8.90

Electronic Configuration	[Ar] $3d^8$, $4s^2$

Oxidation State(s)	– 1,0, +1, $\underline{+2}$, +3, +4, +6	Isotopes	^{58}Ni(68.3%), ^{60}Ni(26.1%)

S.E.P.	$E^\ominus$/Volts
Ni^{2+}|Ni	– 0.26 Acid solution

Electronegativity (Pauling)	1.8

Radii	nm
Ionic	0.078 (+2)
Atomic	0.125
Van der Waals	0.160
Covalent	0.115

Enthalpy Data	kJ mol^{-1}
1st Ionisation Energy	737
2nd Ionisation Energy	1753
1st Electron Affinity	+111.0
2nd Electron Affinity	–
Enthalpy of Fusion, $\Delta H^\ominus_{Fusion}$	17.6
Enthalpy of Vaporisation, $\Delta H^\ominus_{Vaporisation}$	374.8
Enthalpy of Atomisation, $\Delta H^\ominus_{Atomisation}$	429.7
	J $K^{-1}mol^{-1}$
Standard Entropy, $S^\ominus$	29.9

Associated Bond Lengths (nm) & Energies (kJ mol^{-1})	n.a.

Niobium

[Latin, 'Niobe' daughter of Tantalus]

Atomic Number	41
Relative Atomic Mass	92.9064

Chemical Symbol	Nb	Group	–	Period	5 2nd series T.M.

Main Source & History

Discovered by C. Hatchett in 1801, London, UK. The most important mineral is columbite (niobite) ($(Fe,Mn)Nb_2O_6$). The mineral is treated with sodium hydroxide and fused to give Nb_2O_5 which is then reduced to the metal by reduction with carbon.

Properties

Niobium is a lustrous metal, of high melting point. The element is unreactive towards water but reacts with oxygen and steam (liberating hydrogen) at high temperature. It is resistant to acids apart from a mixture of nitric and hydrofluoric acids.

Uses

Niobium is used in the manufacture of stainless steels due to its ability to maintain the steel's resistance to corrosion at high temperatures. Niobium alloys with titanium are superconducting.

Biological Role

Niobium has no known biological role.

Melting Point/K	2740.0	Boiling Point/K	5015.0	Density/g cm^{-3}	8.57^{293K}

Electronic Configuration	[Kr] $4d^4$, $5s^1$

Oxidation State(s)	– 1,0,+1,+2,+3,+4,$\underline{+5}$	Isotopes	^{93}Nb(100%)

S.E.P.	$E^\ominus$/Volts
Nb_2O_5|Nb	−0.65 Acid solution

Electronegativity (Pauling)	1.6

Radii	nm
Ionic	0.070 (+5)
Atomic	0.143
Van der Waals	–
Covalent	0.134

Enthalpy Data	kJ mol^{-1}
1st Ionisation Energy	664
2nd Ionisation Energy	1382
1st Electron Affinity	+109.0
2nd Electron Affinity	–
Enthalpy of Fusion, $\Delta H^\ominus_{Fusion}$	26.8
Enthalpy of Vaporisation, $\Delta H^\ominus_{Vaporisation}$	696.6
Enthalpy of Atomisation, $\Delta H^\ominus_{Atomisation}$	725.9
	J $K^{-1}mol^{-1}$
Standard Entropy, $S^\ominus$	36.4

Associated Bond Lengths (nm) & Energies (kJ mol^{-1})	n.a.

Nitrogen

[Greek, 'nitro genes' = salt petre forming]

Atomic Number	7
Relative Atomic Mass	14.0067

Chemical Symbol	N	Group	5	Period	2

Main Source & History

Discovered by D. Rutherford in 1772, Edinburgh, UK.
It is obtained by the fractional distillation of liquid air.

Properties

It is a colourless, odourless, tastless gas which is slightly less dense than air. The chief characteristic of the gas is its chemical inertness, although it will react with metals such as lithium and magnesium and non–metals such as hydrogen in the production of ammonia.

Uses

It is used as one of the raw materials of the Haber Process for the manufacture of ammonia, as well as in the manufacture of nitric acid and hence fertilisers. Nitrogen is also involved in the production of plastics (nylon), explosives (T.N.T.), dyes (diazo) and as an inert atmosphere in certain metallurgical and chemical processes and as the inflating agent in crisp packaging. Liquid nitrogen is used as a refrigerant as well as for low temperature research work.

Biological Role

Nitrogen is found in the majority of substances which are essential for life, for example, DNA, proteins and porphyrins. A deficiency of nitrogen will be seen in oedema and weakness due to lack of protein as well as in stunted growth. The nitrogen cycle involves atmospheric nitrogen being brought into the life cycle. Nitrogen deficiency in plants gives rise to chlorosis.

Melting Point/K	63.0	Boiling Point/K	77.4	Density/g cm^{-3}	0.88^{77K}

Electronic Configuration	[He] $2s^2\ 2p^3$

Oxidation State(s)	$\underline{-3}$,−2,−1,0,+1,+2,$\underline{\pm 3}$,+4,$\underline{\pm 5}$	Isotopes	^{14}N(99.63%), ^{15}N(0.37%)

S.E.P.	$E^{\ominus}$/Volts
$NO_3^-\,\vert\,{}^1/_2N_2$	+1.24 (Acid solution)

Electronegativity (Pauling)	3.0

Radii	nm
Ionic	0.171 (−3)
Atomic	0.055
Van der Waals	0.154
Covalent	0.074

Enthalpy Data	kJ mol^{-1}
1st Ionisation Energy	1402
2nd Ionisation Energy	2856
1st Electron Affinity	−3.0
2nd Electron Affinity	+800.0
Enthalpy of Fusion, $\Delta H^{\ominus}_{Fusion}$	0.4
Enthalpy of Vaporisation, $\Delta H^{\ominus}_{Vaporisation}$	2.8
Enthalpy of Atomisation, $\Delta H^{\ominus}_{Atomisation}$	472.7
	J $K^{-1}mol^{-1}$
Standard Entropy, $S^{\ominus}$	95.8

Associated Bond Lengths (nm) & Energies (kJ mol^{-1})	N−N(0.147,160), N=N(0.125,415), N≡N(0.110,946), N−H(0.101,390), N−O(0.120,214), N≡C(0.116,890)

Nobelium

[After A.B. Nobel]

Atomic Number	102
Relative Atomic Mass	259.1009[1]

Chemical Symbol	No	Group	–	Period	7 Actinide series

Main Source & History

Discovered by A. Ghiorso, G.T. Seaborg, T. Sikkeland and J.R. Walton in 1958, California, USA. The metal has not yet been prepared, however, a few atoms have been obtained by bombardment of ^{246}Cm with carbon nuclei.

Properties

This transuranic element is a radioactive metal. The most stable isotope ^{259}No has a $T_{1/2}$ of 58 minutes.

Uses	Nobelium has no known uses.

Biological Role	Nobelium has no known biological role. It is hazardous to health due to its radioactive nature.

Melting Point/K	–	Boiling Point/K	–	Density/g cm^{-3}	–

Electronic Configuration	[Rn] $5f^{14}$, $7s^2$

Oxidation State(s)	0,$\underline{\pm 2}$,+3	Isotopes	^{259}No(0%), ^{255}No(0%)

S.E.P.	$E^{\ominus}$/Volts
No^{3+}|No	– 1.78 (Acid solution)

Electronegativity (Pauling)	1.3

Radii	nm
Ionic	0.113 (+2)
Atomic	–
Van der Waals	–
Covalent	–

Enthalpy Data	kJ mol^{-1}
1st Ionisation Energy	642
2nd Ionisation Energy	–
1st Electron Affinity	–
2nd Electron Affinity	–
Enthalpy of Fusion, $\Delta H^{\ominus}_{Fusion}$	–
Enthalpy of Vaporisation, $\Delta H^{\ominus}_{Vaporisation}$	–
Enthalpy of Atomisation, $\Delta H^{\ominus}_{Atomisation}$	–
	J $K^{-1}mol^{-1}$
Standard Entropy, $S^{\ominus}$	–

Associated Bond Lengths (nm) & Energies (kJ mol^{-1})	n.a.

Osmium

[Greek, 'osme' = odour]

Atomic Number	76
Relative Atomic Mass	190.2

Chemical Symbol	Os	Group	–	Period	6 (3rd series T.M.)

Main Source & History

Discovered by S. Tennant in 1803, London, UK. It occurs naturally as a native alloy with Pt, Ir, Rh and Ru called osmiridium. Osmium is oxidised to volatile OsO_4 from which the pure metal is obtained by reduction with hydrogen. It is also obtained as a by–product of nickel refining.

Properties

Osmium is a lustrous, silvery metal. It is the densest of the known elements. This metal is unreactive towards air, water and acids, but will react with fluorine and alkalis at high temperatures.

Uses

Osmium is used in the manufacture of alloys in which it has a hardening effect. With iridium its alloys are used for fountain pen nibs and spark plugs. Osmium and its compounds are used as catalysts, for example, OsO_4 is used to produce stereospecific isomers.

Biological Role

Osmium has no known biological role.

Melting Point/K	3327.0	Boiling Point/K	5300.0	Density/g cm^{-3}	22.48

Electronic Configuration	[Xe] $4f^{14}$, $5d^{6}$, $6s^{2}$

Oxidation State(s)	–2,0,+1,+2,+3,$\underline{+4}$,+5,+6,+7,+8	Isotopes	^{192}Os(41.0%), ^{190}Os(26.4%)

S.E.P.	$E^{\ominus}$/Volts
OsO_2|Os	+ 0.69

Electronegativity (Pauling)	2.2

Radii	nm
Ionic	0.067 (+4)
Atomic	0.134
Van der Waals	–
Covalent	0.126

Enthalpy Data	kJ mol^{-1}
1st Ionisation Energy	840
2nd Ionisation Energy	1630
1st Electron Affinity	+110.0
2nd Electron Affinity	–
Enthalpy of Fusion, $\Delta H^{\ominus}_{Fusion}$	29.3
Enthalpy of Vaporisation, $\Delta H^{\ominus}_{Vaporisation}$	627.6
Enthalpy of Atomisation, $\Delta H^{\ominus}_{Atomisation}$	790.8
	J $K^{-1}mol^{-1}$
Standard Entropy, $S^{\ominus}$	32.6

Associated Bond Lengths (nm) & Energies (kJ mol^{-1})	n.a.

Oxygen

[Greek, 'oxy genes' = acid forming]

Atomic Number	8
Relative Atomic Mass	15.9994

Chemical Symbol	O	Group	6	Period	2

Main Source & History

Discovered by C.W. Scheele (Sweden) and J. Priestly (UK) independently between 1771–1774. It is obtained by the fractional distillation of liquid air.

Properties

Oxygen is a colourless, odourless, tasteless gas. It is slightly denser than air and is appreciably soluble in water. It is a reactive gas and forms oxides with the vast majority of elements.

Uses

Oxygen is used extensively in the manufacture of steel and in the production of many chemicals, for example nitric acid, sulphuric acid and epoxyethane. It is used to produce high temperature flames for metal cutting and welding as well as a component of rocket fuels. Oxygen is also used for breathing apparatus for medical uses as well as for diving.

Biological Role

Oxygen is universal in organic compounds of the cell, for example DNA. It is essential for life to take place through respiration. Oxygen is transported around the body by haemoglobin in the blood.

Melting Point/K	54.8	Boiling Point/K	90.2	Density/g cm^{-3}	1.15^{90K}

Electronic Configuration	[He] $2s^2$ $2p^4$

Oxidation State(s)	<u>−2</u>, −1, 0, +1, +2	Isotopes	^{16}O(99.76%), ^{17}O(0.04%)

S.E.P.	$E^\ominus$/Volts
$H_2O\|{}^1/_2O_2$	+ 1.23 (Acid solution)

Electronegativity (Pauling)	3.5

Radii	nm
Ionic	0.140 (−2)
Atomic	0.060
Van der Waals	0.140
Covalent	0.073

Enthalpy Data	kJ mol^{-1}
1st Ionisation Energy	1314
2nd Ionisation Energy	3388
1st Electron Affinity	−142.0
2nd Electron Affinity	+844
Enthalpy of Fusion, $\Delta H^\ominus_{Fusion}$	0.22
Enthalpy of Vaporisation, $\Delta H^\ominus_{Vaporisation}$	3.4
Enthalpy of Atomisation, $\Delta H^\ominus_{Atomisation}$	249.2
	J $K^{-1}mol^{-1}$
Standard Entropy, $S^\ominus$	102.5

Associated Bond Lengths (nm) & Energies (kJ mol^{-1})	O–O(0.148,146), O=O(0.121,498), O–H(0.096,464), O–C(0.143,358), O=C(0.116,805), O– Si(0.150,374)

Palladium

[After the asteroid Pallas]

Atomic Number	46
Relative Atomic Mass	106.42

Chemical Symbol	Pd	Group	–	Period	5 2nd series T.M.

Main Source & History

Discovered by W.H. Wollaston in 1803, London, UK. Palladium may be extracted in small quantities from gravels and sands containing approximately 1.4% palladium. Alternatively it may be obtained from zinc and copper ores as a by–product.

Properties

Palladium is a silvery–white lustrous metal. It is malleable and ductile. It is a good absorber of hydrogen gas. Palladium is resistant to corrosion but will react with oxidising acids.

Uses

Palladium is used in the manufacture of alloys such as white gold which is used to make jewellery. It is also used as thin sheets for protection against corrosion and for decoration. It has extensive use as a catalyst particularly in the hydrogenation of alkenes.

Biological Role

Palladium has no known biological role.

Melting Point/K	1825.0	Boiling Point/K	3243.0	Density/g cm^{-3}	12.02^{293K}

Electronic Configuration	[Kr] $4d^{10}$

Oxidation State(s)	0, $\underline{\pm 2}$, +4	Isotopes	^{106}Pd(27.3%), ^{108}Pd(26.7%)

S.E.P.	$E^{\ominus}$/Volts
Pd^{2+}|Pd	+ 0.92 Acid solution

Electronegativity (Pauling)	2.2

Radii	nm
Ionic	0.086 (+2)
Atomic	0.138
Van der Waals	–
Covalent	0.128

Enthalpy Data	kJ mol^{-1}
1st Ionisation Energy	805
2nd Ionisation Energy	1875
1st Electron Affinity	+60.0
2nd Electron Affinity	–
Enthalpy of Fusion, $\Delta H^{\ominus}_{Fusion}$	17.0
Enthalpy of Vaporisation, $\Delta H^{\ominus}_{Vaporisation}$	380.0
Enthalpy of Atomisation, $\Delta H^{\ominus}_{Atomisation}$	378.2
	J $K^{-1}mol^{-1}$
Standard Entropy, $S^{\ominus}$	37.6

Associated Bond Lengths (nm) & Energies (kJ mol^{-1})	n.a.

Phosphorus

[Greek, 'phosphoros' = light bringer]

Atomic Number	15
Relative Atomic Mass	30.9738

Chemical Symbol	P	Group	5	Period	3

Main Source & History

Discovered by H. Brand in 1669, Hamburg, Germany. It occurs naturally as calcium phosphate ($Ca_3(PO_4)_2$) and fluoroapatite. It is obtained from calcium phosphate by fusion with coke and sand in an electric furnace. Phosphorus distils out and is condensed under water.

Properties

Phosphorus, a non–metal, has two well defined allotropes. White (P_4) which is a soft, waxy, translucent substance which is extremely reactive (especially towards oxygen), and the macro–molecular red phosphorus which is denser than the white variety, and is relatively unreactive. White phosphorus gradually changes to red.

Uses

The most important use of the element is for the manufacture of matches but it is also used to make alloys (steels and phosphor bronze). Compounds are used in the production of fertilisers (P_2O_5), detergents (Na_3PO_4), insecticides, foods ($CaHPO_4$) and drinks (H_3PO_4), pesticides and special glasses.

Biological Role

Phosphorus is essential to life as the constituent of cell membranes as well as some proteins and all nucleic acids and nucleotides. Calcium phosphate is the main constituent in bones and hence is required for the formation of the skeleton. Phosphorus is required for the phosphorylation of sugar in glycolysis. In plants phosphorus deficiency produces poor growth.

Melting Point/K	317 white 863^{4atm} red	Boiling Point/K	553 white 473 red ignites	Density/g cm^{-3}	1.82 white 2.34 red

Electronic Configuration	[Ne] $3s^2 3p^3$

Oxidation State(s)	<u>−3</u>, −2, 0, +2, <u>±3</u>, <u>±5</u>	Isotopes	^{31}P(100%)

S.E.P.	$E^{\ominus}$/Volts
$P\|PH_3$	− 0.06 (Acid solution)

Electronegativity (Pauling)	2.1

Radii	nm
Ionic	0.017 (+5)
Atomic	0.110
Van der Waals	0.190
Covalent	0.110

Enthalpy Data	kJ mol^{-1}
1st Ionisation Energy	1012
2nd Ionisation Energy	1903
1st Electron Affinity	−70.0
2nd Electron Affinity	–
Enthalpy of Fusion, $\Delta H^{\ominus}_{Fusion}$	0.6^{white}
Enthalpy of Vaporisation, $\Delta H^{\ominus}_{Vaporisation}$	12.4^{white}
Enthalpy of Atomisation, $\Delta H^{\ominus}_{Atomisation}$	314.6
	J $K^{-1} mol^{-1}$
Standard Entropy, $S^{\ominus}$	41.1^{white}

Associated Bond Lengths (nm) & Energies (kJ mol^{-1})	P–P(0.222,209), P–H(0.144,328), P–O(0.164,407), P–Cl(0.204,319)

Platinum

[Spanish, 'platina' = silver]

Atomic Number	78
Relative Atomic Mass	195.08

Chemical Symbol	Pt	Group	–	Period	6 3rd series T.M.

Main Source & History

Discovered by A. de Ulloa in 1748, Columbia, although platinum appears to have been known to South Americans prior to this 'discovery'. Platinum occurs naturally alloyed with osmium, iridium and similar metals. It is formed as a by–product in the extraction of copper and nickel.

Properties

Platinum is a bluish–white, lustrous metal. It is very malleable and ductile. Platinum is relatively unreactive towards oxygen and water but will dissolve in oxidising acids.

Uses	Platinum is used in the manufacture of jewellery, alloys (with Ir and Rh, as thermocouples), and as electrical contacts. It is used extensively as a catalyst, for example in the manufacture of ammonia, nitric acid, sulphuric acid and methanal. It is also used in the cracking of hydrocarbons. Compounds of platinum containing a platinum(II) complex ion are used as anti– tumour drugs.
Biological Role	Platinum has no known biological role.

Melting Point/K	2045.0	Boiling Point/K	4100.0^2	Density/g cm^{-3}	21.45

Electronic Configuration	[Xe] $4f^{14}$, $5d^9$, $6s^1$

Oxidation State(s)	0, <u>±2</u>, <u>±4</u>, +5, +6	Isotopes	^{195}Pt(33.8%), ^{194}Pt(32.9%)

S.E.P.	$E^{\ominus}$/Volts
Pt^{2+}|Pt	+ 1.19

Electronegativity (Pauling)	2.2

Radii	nm
Ionic	0.060 (+2)
Atomic	0.138
Van der Waals	–
Covalent	0.129

Enthalpy Data	kJ mol^{-1}
1st Ionisation Energy	870
2nd Ionisation Energy	1791
1st Electron Affinity	+205.3
2nd Electron Affinity	–
Enthalpy of Fusion, $\Delta H^{\ominus}_{Fusion}$	19.7
Enthalpy of Vaporisation, $\Delta H^{\ominus}_{Vaporisation}$	510.0
Enthalpy of Atomisation, $\Delta H^{\ominus}_{Atomisation}$	565.3
	J $K^{-1}mol^{-1}$
Standard Entropy, $S^{\ominus}$	41.6

Associated Bond Lengths (nm) & Energies (kJ mol^{-1})	n.a.

Plutonium

[After the planet Pluto]

Atomic Number	94
Relative Atomic Mass	244.0642[1]

Chemical Symbol	Pu	Group	–	Period	7 Actinide series

Main Source & History

Discovered by J.W. Kennedy, E.M. McMillan, G.T. Seaborg and A.C. Wahl in 1940, California, USA. It occurs naturally in ores of uranium. It is separated by solvent extraction. The metal is obtained by the reduction of PuF_4 with calcium.

Properties

This transuranic element is a radioactive, silvery metal. The most stable isotope ^{244}Pu has a $T_{1/2}$ of 8.2×10^7 years. Plutonium reacts with oxygen and liberates hydrogen from steam. It also reacts with acids.

Uses

Plutonium is used extensively as a nuclear fuel and for nuclear weapons. ^{238}Pu is also used as a compact energy source for space vehicles. The large quantities required for these uses are obtained from uranium fuel elements.

Biological Role

Plutonium has no known biological role. It is hazardous to health due to its highly radioactive nature.

Melting Point/K	914.0	Boiling Point/K	3505.0	Density/g cm^{-3}	19.84

Electronic Configuration	[Rn] $5f^6, 7s^2$

Oxidation State(s)	0, +2, +3, $\underline{+4}$, +5, +6, +7	Isotopes	^{244}Pu(trace), ^{242}Pu(trace)

S.E.P.	$E^\ominus$/Volts
Pu^{4+}|Pu	– 1.25 (Acid solution)

Electronegativity (Pauling)	1.3

Radii	nm
Ionic	0.093 (+4)
Atomic	0.151
Van der Waals	–
Covalent	–

Enthalpy Data	kJ mol^{-1}
1st Ionisation Energy	585
2nd Ionisation Energy	–
1st Electron Affinity	–
2nd Electron Affinity	–
Enthalpy of Fusion, $\Delta H^\ominus_{Fusion}$	2.8
Enthalpy of Vaporisation, $\Delta H^\ominus_{Vaporisation}$	343.5
Enthalpy of Atomisation, $\Delta H^\ominus_{Atomisation}$	–
	J $K^{-1}mol^{-1}$
Standard Entropy, $S^\ominus$	–

Associated Bond Lengths (nm) & Energies (kJ mol^{-1})	n.a.

Polonium

[After Poland]

Atomic Number	84
Relative Atomic Mass	208.9824[1]

Chemical Symbol	Po	Group	6	Period	6

Main Source & History

Discovered by Mme M.S. Curie in 1898, Paris, France. It is obtained in gram quantities from the neutron irradiation of ^{209}Bi. It can be separated from bismuth by electrochemical methods and purified by sublimation.

Properties

Polonium is a soft, silvery–grey, radioactive metal. The most stable isotope ^{209}Po has a $T_{1/2}$ of 103 years.

Uses

Polonium is used as an alpha radiation source.

Biological Role

Polonium has no known biological role. It is hazardous to health due to its highly radioactive nature.

Melting Point/K	527.0	Boiling Point/K	1235.0	Density/g cm^{-3}	9.40

Electronic Configuration	[Xe] $4f^{14}$, $5d^{10}$, $6s^2$ $6p^4$

Oxidation State(s)	– 2,0,+2,<u>+4</u>,+6	Isotopes	^{210}Po(trace), ^{211}Po(trace)

S.E.P.	$E^{\ominus}$/Volts
PoO_2|Po	+ 0.73 (Acid Solution)

Electronegativity (Pauling)	2.0

Radii	nm
Ionic	0.064 (+4)
Atomic	0.167
Van der Waals	–
Covalent	0.153

Enthalpy Data	kJ mol^{-1}
1st Ionisation Energy	812
2nd Ionisation Energy	1800[2]
1st Electron Affinity	–180.0
2nd Electron Affinity	–
Enthalpy of Fusion, $\Delta H^{\ominus}_{Fusion}$	12.6
Enthalpy of Vaporisation, $\Delta H^{\ominus}_{Vaporisation}$	100.8
Enthalpy of Atomisation, $\Delta H^{\ominus}_{Atomisation}$	144.1
	J $K^{-1}mol^{-1}$
Standard Entropy, $S^{\ominus}$	62.8

Associated Bond Lengths (nm) & Energies (kJ mol^{-1})	n.a.

Potassium

[English, potash]

Atomic Number	19
Relative Atomic Mass	39.0983

Chemical Symbol	K	Group	1	Period	4

Main Source & History

Discovered by Sir H. Davy in 1807, London, UK. It occurs principally as the ore carnalite ($KMgCl_3.6H_2O$) but also as sylvite (KCl) and salt petre (KNO_3). It is extracted by the electrolysis of the fused halide.

Properties

Potassium is a soft, silvery–white metal of low density. The element rapidly tarnishes when exposed to the air, reacting vigorously with oxygen and water from which it liberates hydrogen.

Uses

Potassium is used, when mixed with sodium, as a heat transfer agent in nuclear power stations. It is also used as a reducing agent. Compounds of potassium are used in fertilisers (KCl), glasses (KOH), batteries (KOH), medicines (KBr) and the paint and dye industry ($K_2Cr_2O_7$), as well as for food additives ($KHCO_3$), explosives (KNO_3) and soaps (KOH).

Biological Role

Potassium is one of the principal minerals in the body. It is present in body fluids as the potassium ion (K^+). This ion is involved in homeostasis and is important to the functioning of the central nervous system, protein synthesis and muscle contraction. K^+ is also involved in the action of the enzyme tryptophanase. In plants it is involved in the formation of cell membranes and the opening of stomata.

Melting Point/K	337.0	Boiling Point/K	1047.0	Density/g cm^{-3}	0.86

Electronic Configuration	[Ar] $4s^1$

Oxidation State(s)	−1,0,±1	Isotopes	^{39}K(93.26%), ^{41}K(6.73%)

S.E.P.	$E^\ominus$/Volts
K^+|K	− 2.93

Electronegativity (Pauling)	0.8

Radii	nm
Ionic	0.133 (+1)
Atomic	0.227
Van der Waals	0.231
Covalent	0.203

Enthalpy Data	kJ mol^{-1}
1st Ionisation Energy	419
2nd Ionisation Energy	3051
1st Electron Affinity	−48.4
2nd Electron Affinity	–
Enthalpy of Fusion, $\Delta H^\ominus_{Fusion}$	2.3
Enthalpy of Vaporisation, $\Delta H^\ominus_{Vaporisation}$	79.1
Enthalpy of Atomisation, $\Delta H^\ominus_{Atomisation}$	89.2
	J $K^{-1}mol^{-1}$
Standard Entropy, $S^\ominus$	64.2

Associated Bond Lengths (nm) & Energies (kJ mol^{-1})	n.a.

Praseodymium

[Greek, 'prasios didymos' = green twin]

Atomic Number	59
Relative Atomic Mass	140.9077

Chemical Symbol	Pr	Group	–	Period	6 Lanthanide series

Main Source & History

Discovered by Baron A. von Welsbach in 1885, Vienna, Austria. The main ores are monazite and bastnaesite from which praseodymium is extracted, with difficulty, by ion exchange. The metal can be obtained by the reduction of the oxide.

Properties

Praseodymium is a lustrous metal. It is a good conductor of heat and electricity and is quite soft and malleable. Praseodymium is quite a reactive metal. It reacts with water liberating hydrogen, but only slowly on exposure to air. Praseodymium also reacts with acids.

Uses

Praseodymium is used to make special glasses and alloys (with iron) from which 'flints' are made.

Biological Role

Praseodymium has no known biological role.

Melting Point/K	1204.0	Boiling Point/K	3785.0	Density/g cm^{-3}	6.48^{293K}

Electronic Configuration	[Xe] $4f^3$, $6s^2$

Oxidation State(s)	0,$\underline{+3}$,+4	Isotopes	^{141}Pr(100%)

S.E.P.	$E^{\ominus}$/Volts
Pr^{3+}|Pr	– 2.35 (Acid solution)

Electronegativity (Pauling)	1.1

Radii	nm
Ionic	0.106 (+3)
Atomic	0.183
Van der Waals	–
Covalent	0.165

Enthalpy Data	kJ mol^{-1}
1st Ionisation Energy	523
2nd Ionisation Energy	1018
1st Electron Affinity	+50.0[2]
2nd Electron Affinity	–
Enthalpy of Fusion, $\Delta H^{\ominus}_{Fusion}$	11.3
Enthalpy of Vaporisation, $\Delta H^{\ominus}_{Vaporisation}$	357.0
Enthalpy of Atomisation, $\Delta H^{\ominus}_{Atomisation}$	–
	J $K^{-1}mol^{-1}$
Standard Entropy, $S^{\ominus}$	73.2

Associated Bond Lengths (nm) & Energies (kJ mol^{-1})	n.a.

Promethium

[After Greek god Promethius]

Atomic Number	61
Relative Atomic Mass	146.9151[1]

Chemical Symbol	Pm	Group	–	Period	6 Lanthanide series

Main Source & History

Discovered by C.D. Coryell, L.E. Glendenin and G.A. Marinsky in 1945, California, USA. Promethium occurs in small quantities in uranium ores. It is however extracted in milligram quantities from the fission products of uranium in nuclear reactors.

Properties

Promethium is a radioactive metal. The most stable isotope ^{145}Pm has a $T_{1/2}$ of 17.7 years. Promethium reacts with air, and water from which it liberates hydrogen. It also reacts with acids.

Uses

Promethium has found uses in the manufacture of miniature batteries and in luminescent paint.

Biological Role

Promethium has no known biological role. It is hazardous to health due to its radioactive nature.

Melting Point/K	1441.0	Boiling Point/K	3000.0[2]	Density/g cm^{-3}	7.22[2]

Electronic Configuration	[Xe] $4f^5$, $6s^2$

Oxidation State(s)	0, $\underline{+3}$	Isotopes	^{145}Pm(trace), ^{147}Pm(0%)

S.E.P.	$E^{\ominus}$/Volts
Pm^{3+}|Pm	– 2.29 (Acid solution)

Electronegativity (Pauling)	1.2

Radii	nm
Ionic	0.106 (+3)
Atomic	0.181
Van der Waals	–
Covalent	0.163

Enthalpy Data	kJ mol^{-1}
1st Ionisation Energy	536
2nd Ionisation Energy	1052
1st Electron Affinity	+50.0[2]
2nd Electron Affinity	–
Enthalpy of Fusion, $\Delta H^{\ominus}_{Fusion}$	12.6
Enthalpy of Vaporisation, $\Delta H^{\ominus}_{Vaporisation}$	–
Enthalpy of Atomisation, $\Delta H^{\ominus}_{Atomisation}$	–
	J $K^{-1}mol^{-1}$
Standard Entropy, $S^{\ominus}$	–

Associated Bond Lengths (nm) & Energies (kJ mol^{-1})	n.a.

Protactinium

[Greek, 'protos' = first]

Atomic Number	91
Relative Atomic Mass	231.0359[1]

Chemical Symbol	Pa	Group	–	Period	7 Actinide series

Main Source & History

Discovered by J.A. Cranston, A. Fleck and F. Soddy (UK), O. Hahn and L. Meitner (Germany) independently in 1917. It is separated from residues from uranium production by solvent extraction. The metal can be obtained by reduction of PaF_4 with barium.

Properties

Protactinium is a radioactive, lustrous metal. The most stable isotope ^{231}Pa has a $T_{1/2}$ of 3.26×10^4 years. It is a quite reactive metal reacting with oxygen as well as steam from which it liberates hydrogen. It also reacts with acids.

Uses

Protactinium has no known uses.

Biological Role

Protactinium has no known biological role. It is hazardous to health due to its radioactive nature.

Melting Point/K	1872.0[2]	Boiling Point/K	4300.0[2]	Density/g cm^{-3}	15.37[2]

Electronic Configuration	[Rn] $5f^2$, $6d^1$, $7s^2$

Oxidation State(s)	0, +3, +4, <u>+5</u>	Isotopes	^{231}Pa(trace), ^{234}Pa(trace)

S.E.P.	$E^\ominus$/Volts
Pa^{5+}|Pa	– 1.20 (Acid Solution)

Electronegativity (Pauling)	1.5

Radii	nm
Ionic	0.089(+5)
Atomic	0.161
Van der Waals	–
Covalent	–

Enthalpy Data	kJ mol^{-1}
1st Ionisation Energy	568
2nd Ionisation Energy	–
1st Electron Affinity	–
2nd Electron Affinity	–
Enthalpy of Fusion, $\Delta H^\ominus_{Fusion}$	14.6
Enthalpy of Vaporisation, $\Delta H^\ominus_{Vaporisation}$	460.2
Enthalpy of Atomisation, $\Delta H^\ominus_{Atomisation}$	606.7
	J $K^{-1}mol^{-1}$
Standard Entropy, $S^\ominus$	51.9

Associated Bond Lengths (nm) & Energies (kJ mol^{-1})	n.a.

Radium

[Latin, 'radius' = ray]

Atomic Number	88
Relative Atomic Mass	226.0254[1]

Chemical Symbol	Ra	Group	2	Period	7

Main Source & History

Discovered by M.G. Bemont, Mme M.S. Curie and P. Curie in 1898, Paris, France. Radium occurs in uranium ores from which it is separated by electrolysis in gram quantities.

Properties

Radium is a soft, silvery– white, radioactive metal. The most stable isotope ^{226}Ra has a $T_{1/2}$ of 1600 years. Radium reacts with water releasing hydrogen and tarnishes in air.

Uses

Radium is used in radiotherapy in the treatment of cancer. It is also used in some luminous paints.

Biological Role

Radium has no known biological role. It is hazardous to health due to its radioactive nature.

Melting Point/K	973.0	Boiling Point/K	1413.0[2]	Density/g cm^{-3}	5.00[2]

Electronic Configuration	[Rn] $7s^2$

Oxidation State(s)	0,±2	Isotopes	^{226}Ra(trace), ^{228}Ra(trace)

S.E.P.	$E^{\ominus}$/Volts
Ra^{2+}|Ra	– 2.92

Electronegativity (Pauling)	0.9

Radii	nm
Ionic	0.148(+2)
Atomic	0.223
Van der Waals	–
Covalent	–

Enthalpy Data	kJ mol^{-1}
1st Ionisation Energy	510
2nd Ionisation Energy	979
1st Electron Affinity	–
2nd Electron Affinity	–
Enthalpy of Fusion, $\Delta H^{\ominus}_{Fusion}$	8.4
Enthalpy of Vaporisation, $\Delta H^{\ominus}_{Vaporisation}$	136.7
Enthalpy of Atomisation, $\Delta H^{\ominus}_{Atomisation}$	161.9
	J $K^{-1}mol^{-1}$
Standard Entropy, $S^{\ominus}$	71.1

Associated Bond Lengths (nm) & Energies (kJ mol^{-1})	n.a.

Radon

[After the element radium]

Atomic Number	86
Relative Atomic Mass	222.0176[1]

Chemical Symbol	Rn	Group	0 (8)	Period	6

Main Source & History

Discovered by F.E. Dorn in 1900, Germany. It occurs naturally as a decay product of ^{226}Ra.

Properties

Radon is a colourless, odourless, radioactive gas. The most stable isotope ^{222}Rn has a $T_{1/2}$ of 3.8 days. Radon is expected to be the most reactive of the noble gases.

Uses	Radon is used as a source of radiation and as a radioactive gaseous tracer.

Biological Role	Radon has no known biological role. It is hazardous to health due to its radioactive nature.

Melting Point/K	202.0	Boiling Point/K	211.0	Density/g cm^{-3}	4.40^{211K} [2]

Electronic Configuration	[Xe] $4f^{14}$, $5d^{10}$, $6s^2$ $6p^6$

Oxidation State(s)	0,±2	Isotopes	^{222}Rn(trace), ^{220}Rn(trace)

S.E.P. $E^{\ominus}$/Volts
n.a.

Electronegativity (Pauling)	–

Radii	nm
Ionic	–
Atomic	–
Van der Waals	–
Covalent	–

Enthalpy Data	kJ mol^{-1}
1st Ionisation Energy	1037
2nd Ionisation Energy	1930
1st Electron Affinity	0.0
2nd Electron Affinity	–
Enthalpy of Fusion, $\Delta H^{\ominus}_{Fusion}$	2.9
Enthalpy of Vaporisation, $\Delta H^{\ominus}_{Vaporisation}$	16.4
Enthalpy of Atomisation, $\Delta H^{\ominus}_{Atomisation}$	0.0
	J K^{-1} mol^{-1}
Standard Entropy, $S^{\ominus}$	176.1

Associated Bond Lengths (nm) & Energies (kJ mol^{-1})	n.a.

Rhenium

[Latin, 'Rhenus' = the river Rhine]

Atomic Number	75
Relative Atomic Mass	186.207

Chemical Symbol	Re	Group	–	Period	6 3rd series T.M.

Main Source & History	Properties
Discovered by O. Berg, W. Naddack and I. Tacke in 1925, Berlin, Germany. Rhenium compounds obtained from the flue dusts of molybdenum smelters are reduced to the metal using hydrogen.	Rhenium is a whitish–grey metal. It is hard and has a high melting point. It is a relatively unreactive metal reacting slowly with both air and water from which it liberates hydrogen.

Uses	Rhenium is used in the manufacture of alloys which are used in thermocouples (with tungsten), and as filaments. Rhenium is also used in catalysis.

Biological Role	Rhenium has no known biological role.

Melting Point/K	3453.0	Boiling Point/K	5900.0[2]	Density/g cm^{-3}	21.03^{293K}

Electronic Configuration	[Xe] $4f^{14}$, $5d^{5}$, $6s^{2}$

Oxidation State(s)	–1,0,+1,+2,+3,+4,+5,+6,+7	Isotopes	^{187}Re(62.93%), ^{185}Re(37.07%)

S.E.P.	E$^{\ominus}$/Volts
ReO_4^-|Re	+ 0.34 Acid solution

Electronegativity (Pauling)	1.9

Radii	nm
Ionic	0.053 (+7)
Atomic	0.137
Van der Waals	–
Covalent	0.128

Enthalpy Data	kJ mol^{-1}
1st Ionisation Energy	760
2nd Ionisation Energy	1600
1st Electron Affinity	+15.0
2nd Electron Affinity	–
Enthalpy of Fusion, $\Delta H^{\ominus}_{Fusion}$	33.1
Enthalpy of Vaporisation, $\Delta H^{\ominus}_{Vaporisation}$	704.3
Enthalpy of Atomisation, $\Delta H^{\ominus}_{Atomisation}$	769.9
	J $K^{-1}mol^{-1}$
Standard Entropy, $S^{\ominus}$	36.9

Associated Bond Lengths (nm) & Energies (kJ mol^{-1})	n.a.

Rhodium

[Greek, 'rhodon' = rose]

Atomic Number	45
Relative Atomic Mass	102.9055

Chemical Symbol	Rh	Group	–	Period	5 (2nd series T.M.)

Main Source & History

Discovered by W.H. Wollaston in 1804, London, UK. It is found with nickel and copper, rhodium being obtained as a by–product of their extraction. A multi–stage acid extraction process results in rhodium being obtained by reduction of the final compound with hydrogen.

Properties

Rhodium is a hard, steel–blue, lustrous metal. It is relatively chemically inert. Rhodium is not attacked even by acids.

Uses

Rhodium is used mainly in the production of alloys with palladium and platinum which are used in thermocouples and as catalysts, for example in the production of nitric acid. Compounds of rhodium are also used as catalysts ($Rh((C_6H_5)_3P)_3Cl$).

Biological Role

Rhodium has no known biological role.

Melting Point/K	2239.0	Boiling Point/K	4800.0[2]	Density/g cm^{-3}	12.41[293K]

Electronic Configuration	[Kr] $4d^8 5s^1$

Oxidation State(s)	−1,0,+1,+2,<u>+3</u>,+4,+5,+6	Isotopes	^{103}Rh(100%)

S.E.P.	$E^{\ominus}$/Volts
Rh^{3+}|Rh	+ 0.76

Electronegativity (Pauling)	2.3

Radii	nm
Ionic	0.075 (+3)
Atomic	0.135
Van der Waals	–
Covalent	0.125

Enthalpy Data	kJ mol^{-1}
1st Ionisation Energy	720
2nd Ionisation Energy	1745
1st Electron Affinity	+162.0
2nd Electron Affinity	–
Enthalpy of Fusion, $\Delta H^{\ominus}_{Fusion}$	21.6
Enthalpy of Vaporisation, $\Delta H^{\ominus}_{Vaporisation}$	494.3
Enthalpy of Atomisation, $\Delta H^{\ominus}_{Atomisation}$	556.9
	J $K^{-1}mol^{-1}$
Standard Entropy, $S^{\ominus}$	31.5

Associated Bond Lengths (nm) & Energies (kJ mol^{-1})	n.a.

Rubidium

[Latin, 'rubidius' = red]

Atomic Number	37
Relative Atomic Mass	85.4678

Chemical Symbol	Rb	Group	1	Period	5

Main Source & History

Discovered by R.W. Bunsen and G.R. Kirchoff in 1861, Germany. It occurs in the minerals carnalite, pollucite ($Cs(AlSi_2O_6).xH_2O$), lepiodolite and rock salt (NaCl). The metal is obtained by the reduction of RbCl with calcium at high temperature.

Properties

It is an extremely soft, silvery–white metal. Rubidium is chemically extremely reactive, giving vigorous reactions with air as well as water and acids from which it releases hydrogen.

Uses

Rubidium is used in Rb–Sr dating a method for dating rocks which are greater than 1×10^8 years old. It is also used in the production of photo–cells.

Biological Role

Rubidium has no known biological role.

Melting Point/K	312.2	Boiling Point/K	961.0	Density/g cm^{-3}	1.53^{293K}

Electronic Configuration	$[Kr]\ 5s^1$

Oxidation State(s)	−1,0,±1	Isotopes	$^{85}Rb(72.15\%)$,$^{87}Rb(27.85\%)$

S.E.P.	$E^{\ominus}$/Volts
Rb^+|Rb	− 2.93

Electronegativity (Pauling)	0.8

Radii	nm
Ionic	0.149 (+1)
Atomic	0.248
Van der Waals	0.244
Covalent	0.216

Enthalpy Data	kJ mol^{-1}
1st Ionisation Energy	403
2nd Ionisation Energy	2632
1st Electron Affinity	−46.9
2nd Electron Affinity	–
Enthalpy of Fusion, $\Delta H^{\ominus}_{Fusion}$	2.3
Enthalpy of Vaporisation, $\Delta H^{\ominus}_{Vaporisation}$	69.2
Enthalpy of Atomisation, $\Delta H^{\ominus}_{Atomisation}$	80.9
	J $K^{-1} mol^{-1}$
Standard Entropy, $S^{\ominus}$	76.8

Associated Bond Lengths (nm) & Energies (kJ mol^{-1})	n.a.

Ruthenium

[Latin, 'Ruthenia' = Ukraine, USSR]

Atomic Number	44
Relative Atomic Mass	101.07

Chemical Symbol	Ru	Group	–	Period	5 (2nd series T.M.)

Main Source & History

Discovered by K. Klaus in 1845, Russia. Ruthenium occurs naturally alloyed with lead, iridium, rhenium and osmium in osmiridium. After electrolysis, ion exchange and solvent extraction ruthenium is obtained from the resulting compound by reduction with hydrogen. It is also obtained as a by–product in the extraction of nickel.

Properties

Ruthenium is a hard, bluish–white, brittle metal. It is a relatively unreactive metal and is not attacked by acids, air or water.

Uses

Ruthenium is used in the manufacture of alloys with platinum and palladium where it has a hardening effect. Compounds of ruthenium are used as catalysts, its hydride–phosphine complexes being used as catalysts in hydrogenation reactions.

Biological Role

Ruthenium has no known biological role.

Melting Point/K	2583.0	Boiling Point/K	4173.0	Density/g cm^{-3}	12.30

Electronic Configuration	[Kr] $4d^7$, $5s^1$

Oxidation State(s)	–2,0,+1,+2,<u>+3</u>,+4,+5,+6,+7,+8	Isotopes	^{102}Ru(31.6%), ^{104}Ru(18.87%)

S.E.P.	$E^{\ominus}$/Volts
RuO_2|Ru	+ 0.68

Electronegativity (Pauling)	2.2

Radii	nm
Ionic	0.068 (+3)
Atomic	0.133
Van der Waals	–
Covalent	0.124

Enthalpy Data	kJ mol^{-1}
1st Ionisation Energy	711
2nd Ionisation Energy	1617
1st Electron Affinity	+110.0
2nd Electron Affinity	–
Enthalpy of Fusion, $\Delta H^{\ominus}_{Fusion}$	25.5
Enthalpy of Vaporisation, $\Delta H^{\ominus}_{Vaporisation}$	567.0
Enthalpy of Atomisation, $\Delta H^{\ominus}_{Atomisation}$	642.7
	J $K^{-1}mol^{-1}$
Standard Entropy, $S^{\ominus}$	28.5

Associated Bond Lengths (nm) & Energies (kJ mol^{-1})	n.a.

Samarium

[After the mineral samarskite]

Atomic Number	62
Relative Atomic Mass	150.36

Chemical Symbol	Sm	Group	–	Period	6 Lanthanide series

Main Source & History

Discovered by L. de Boisbaudran in 1879, Paris, France. The main ore is monazite from which samarium is extracted, with difficulty, by ion exchange.

Properties

Samarium is a silvery–white element which is a good conductor of heat and electricity. It is a relatively unreactive metal tarnishing only slowly in moist air. Samarium reacts slowly with water liberating hydrogen. It also reacts with acids.

Uses

Samarium is used in the manufacture of alloys which are used as permanent magnets (Sm,Co_5). The compounds of samarium are used in the production of ceramics and special glasses (Sm_2O_3).

Biological Role

Samarium has no known biological role.

Melting Point/K	1350.0	Boiling Point/K	2064.0	Density/g cm^{-3}	7.52^{293K}

Electronic Configuration	[Xe] $4f^6$, $6s^2$

Oxidation State(s)	0, +2, $\underline{+3}$	Isotopes	^{152}Sm(26.6%), ^{154}Sm(22.6%)

S.E.P.	$E^{\ominus}$/Volts
Sm^{3+}|Sm	– 2.30 (Acid solution)

Electronegativity (Pauling)	1.2

Radii	nm
Ionic	0.104 (+3)
Atomic	0.180
Van der Waals	–
Covalent	0.166

Enthalpy Data	$kJ\ mol^{-1}$
1st Ionisation Energy	543
2nd Ionisation Energy	1068
1st Electron Affinity	$+50.0^2$
2nd Electron Affinity	–
Enthalpy of Fusion, $\Delta H^{\ominus}_{Fusion}$	10.9
Enthalpy of Vaporisation, $\Delta H^{\ominus}_{Vaporisation}$	164.8
Enthalpy of Atomisation, $\Delta H^{\ominus}_{Atomisation}$	–
	$J\ K^{-1} mol^{-1}$
Standard Entropy, $S^{\ominus}$	69.6

Associated Bond Lengths (nm) & Energies ($kJ\ mol^{-1}$)	n.a.

Scandium

[Latin, 'Scandia' = Scandanavia]

Atomic Number	21
Relative Atomic Mass	44.9559

Chemical Symbol	Sc	Group	–	Period	4 (1st series T.M.)

Main Source & History

Discovered by L.F. Nilson in 1879, Uppsala, Sweden. The main ore of scandium is thortveitite ($Sc_2Si_2O_7$), it also occurs in monazite. The ore is converted to ScF_3 from which scandium is obtained by reduction with calcium.

Properties

Scandium is a silvery–white metal. It is a relatively reactive element having reactions similar to aluminium.

Uses

The metal scandium is little used, however, its compounds have a variety of uses such as electronic components (Sc_2O_3), high intensity bulbs (ScI_3) and as a hardening agent (ScC) for titanium carbide to make it the second hardest substance known.

Biological Role

Scandium has no known biological role.

Melting Point/K	1814.0	Boiling Point/K	3104.0	Density/g cm^{-3}	2.99

Electronic Configuration	[Ar] $3d^1$, $4s^2$

Oxidation State(s)	0,+3	Isotopes	^{45}Sc(100%)

S.E.P.	$E^\ominus$/Volts
Sc^{3+}|Sc	– 2.03

Electronegativity (Pauling)	1.3

Radii	nm
Ionic	0.075 (+3)
Atomic	0.161
Van der Waals	–
Covalent	0.144

Enthalpy Data	kJ mol^{-1}
1st Ionisation Energy	631
2nd Ionisation Energy	1235
1st Electron Affinity	+70.0
2nd Electron Affinity	–
Enthalpy of Fusion, $\Delta H^\ominus_{Fusion}$	15.9
Enthalpy of Vaporisation, $\Delta H^\ominus_{Vaporisation}$	310.0
Enthalpy of Atomisation, $\Delta H^\ominus_{Atomisation}$	377.8
	J $K^{-1}mol^{-1}$
Standard Entropy, $S^\ominus$	34.6

Associated Bond Lengths (nm) & Energies (kJ mol^{-1})	n.a.

Selenium

[Greek, 'selene' = moon]

Atomic Number	34
Relative Atomic Mass	78.96

Chemical Symbol	Se	Group	6	Period	4

Main Source & History

Discovered by J.J. Berzelius in 1817, Stockholm, Sweden. It is found in the ores clausthalite (PbSe) and crookesite $((CuTlAg)_2Se)$. It is also recovered from the flue dusts where sulphide ores of zinc and copper have been used. The process of extraction involves several stages culminating in the precipitation of scandium with sulphur dioxide.

Properties

Selenium has three allotropes, two of which are 'non–metallic' in nature whilst the other is the 'metallic' form. The metallic allotrope is silvery–grey and is only weakly conducting. Selenium forms compounds with many elements. It also reacts with alkalis and acids.

Uses

Selenium is used in photoelectric cells due to its enhanced conductivity in the presence of light. It is also used in xerography (photocopiers) as well as in the production of steel alloys, semiconductors and solar cells.

Biological Role

Selenium is an essential trace element being associated with vitamin E (tocopherol). This vitamin is believed to play a vital part in the transfer of hydrogen atoms in cell respiration. However, it should be noted that selenium is quite toxic if the necessary requirement is exceeded.

Melting Point/K	490.0	Boiling Point/K	958.1	Density/g cm^{-3}	4.81

Electronic Configuration	[Ar] $3d^{10}$, $4s^2$ $4p^4$

Oxidation State(s)	– 2,0,+1,±4,±6	Isotopes	$^{80}Se(49.6\%)$, $^{78}Se(23.5\%)$

S.E.P.	$E^{\ominus}$/Volts
H_2SeO_3|Se	+0.74 (Acid solution)

Electronegativity (Pauling)	2.4

Radii	nm
Ionic	0.029 (+6)
Atomic	0.116
Van der Waals	0.200
Covalent	0.117

Enthalpy Data	kJ mol^{-1}
1st Ionisation Energy	941
2nd Ionisation Energy	2045
1st Electron Affinity	−195.0
2nd Electron Affinity	+420.0
Enthalpy of Fusion, $\Delta H^{\ominus}_{Fusion}$	5.2
Enthalpy of Vaporisation, $\Delta H^{\ominus}_{Vaporisation}$	26.3
Enthalpy of Atomisation, $\Delta H^{\ominus}_{Atomisation}$	227.1
	J $K^{-1}mol^{-1}$
Standard Entropy, $S^{\ominus}$	42.4

Associated Bond Lengths (nm) & Energies (kJ mol^{-1})	Se–Se(0.232,330), Se–H(0.146,313), Se–O(0.161,343), Se–Cl(0.220,245)

Silicon

[Latin, 'silex' = flint]

Atomic Number	14
Relative Atomic Mass	28.0855

Chemical Symbol	Si	Group	4	Period	3

Main Source & History

Discovered by J.J. Berzelius in 1824, Stockholm, Sweden. Silicon is the second most abundant element in the earth's crust. It is found as silica (SiO_2) and silicates. Silicon is obtained by reduction with carbon in an electric furnace. Very pure silicon is obatined by zone refining.

Properties

Pure silicon is a dark–grey solid with the same structure as diamond. It has a high melting point and is a semiconductor. Powdered silicon is not attacked by acids but is attacked by the halogens forming halides.

Uses

Silicon is used to make microelectronic devices. Silicates are used in glass manufacture (borosilicates), cements, adhesives (Na_2SiO_3) and detergents. Silicones are widely used to make synthetic rubbers ($(CH_3)_2SiO$) as well as lubricants, for water repellant finishes, varnishes, polishes and paints.

Biological Role

Silicon has no known biological role in humans, however, it is important to lower life forms.

Melting Point/K	1683.0	Boiling Point/K	2628.0	Density/g cm^{-3}	2.33

Electronic Configuration	[Ne] $3s^2$ $3p^2$

Oxidation State(s)	−4,0,+2,±4	Isotopes	^{28}Si(92.23%), ^{29}Si(4.67%)

S.E.P.	$E^{\ominus}$/Volts
SiH_4|Si	−0.14 (Acid solution)

Electronegativity (Pauling)	1.8

Radii	nm
Ionic	0.041 (+4)
Atomic	0.117
Van der Waals	0.210
Covalent	0.117

Enthalpy Data	kJ mol^{-1}
1st Ionisation Energy	787
2nd Ionisation Energy	1577
1st Electron Affinity	−144.7
2nd Electron Affinity	–
Enthalpy of Fusion, $\Delta H^{\ominus}_{Fusion}$	46.4
Enthalpy of Vaporisation, $\Delta H^{\ominus}_{Vaporisation}$	376.8
Enthalpy of Atomisation, $\Delta H^{\ominus}_{Atomisation}$	455.6
	J $K^{-1}mol^{-1}$
Standard Entropy, $S^{\ominus}$	18.8

Associated Bond Lengths (nm) & Energies (kJ mol^{-1})	Si–Si(0.235,226), Si–H(0.148,326), Si–O(0.150,374), Si–Cl(0.202,391)

Silver

[Anglo–Saxon, 'siolfur' = silver]

Atomic Number	47
Relative Atomic Mass	107.8682

Chemical Symbol	Ag	Group	–	Period	5 (2nd series T.M.)

Main Source & History

Silver has been known since pre–historic times. It is found native, however, it occurs in ores such as argentite (Ag_2S). It may be extracted as a cyanide complex from its ore, reduced with zinc and subsequently purified electrolytically.

Properties

Silver is a white, lustrous metal. It is soft, malleable and ductile in nature and has high electrical and thermal conductivity. Silver resists attack by air and water as well as non–oxidising acids.

Uses	Silver has many uses such as jewellery, electronics, silverware, alloying (coinage metals), as a catalyst (formation of alkene oxides) and it even used as a food additive. Silver compounds such as silver iodide are used extensively in photography, antiseptics and for producing 'artificial' rain by the seeding of clouds.

Biological Role	Silver has no known biological role.

Melting Point/K	1235.0	Boiling Point/K	2485.0	Density/g cm^{-3}	10.50

Electronic Configuration	[Kr] $4d^{10}$, $5s^1$

Oxidation State(s)	0,±1,+2,+3	Isotopes	^{107}Ag(51.35%), ^{109}Ag(48.65%)

S.E.P.	$E^\ominus$/Volts
Ag^+|Ag	+0.80 (Acid solution)

Electronegativity (Pauling)	1.9

Radii	nm
Ionic	0.113 (+1)
Atomic	0.145
Van der Waals	0.160
Covalent	0.134

Enthalpy Data	kJ mol^{-1}
1st Ionisation Energy	731
2nd Ionisation Energy	2073
1st Electron Affinity	+125.7
2nd Electron Affinity	–
Enthalpy of Fusion, $\Delta H^\ominus_{Fusion}$	11.3
Enthalpy of Vaporisation, $\Delta H^\ominus_{Vaporisation}$	255.1
Enthalpy of Atomisation, $\Delta H^\ominus_{Atomisation}$	284.6
	J $K^{-1}mol^{-1}$
Standard Entropy, $S^\ominus$	42.6

Associated Bond Lengths (nm) & Energies (kJ mol^{-1})	n.a.

Sodium

[English, soda]

Atomic Number	11
Relative Atomic Mass	22.9898

Chemical Symbol	Na	Group	1	Period	3

Main Source & History

Initially isolated by Sir H. Davy in 1807, London, UK. Obtained commercially by the electrolysis of fused sodium chloride in the Downs Process. This substance occurs naturally as rock salt.

Properties

Sodium is a soft, silvery–white metal. It is an extremely good conductor of electricity and heat. It has a low density and melting point. Because it tarnishes in air it is stored under oil. It reacts vigorously with oxygen as well as water and acids from which it liberates hydrogen.

Uses

The element is used in the preparation of organic compounds such as tetraethyl lead, an anti–knock additive for petroleum (although to a decreasing extent). Sodium is used as a coolant in some nuclear reactors, it is also used in street lighting. Many sodium compounds are used as food additives eg. $NaNO_2$ which prevents the growth of microorganisms. NaCl is used for manufacture of NaOH and Cl_2. It is also used as a preservative and an agent to remove ice and snow.

Biological Role

It is an essential and non–toxic element. Sodium ions (Na^+) play a key role in the process of osmoregulation, which is an example of homeostasis. They, along with potassium and chloride ions, are also necessary for the functioning of the central nervous system. The daily requirement is 4–8 mg. In plants it is essential in halophytes (plants which live on mud flats and salt marshes).

Melting Point/K	370.9	Boiling Point/K	1156.1	Density/g cm^{-3}	0.97

Electronic Configuration	[Ne] $3s^1$

Oxidation State(s)	−1,0,±1	Isotopes	^{23}Na(100%)

S.E.P.	$E^{\ominus}$/Volts
Na^+|Na	− 2.71

Electronegativity (Pauling)	0.9

Radii	nm
Ionic	0.098 (+1)
Atomic	0.186
Van der Waals	0.231
Covalent	0.157

Enthalpy Data	kJ mol^{-1}
1st Ionisation Energy	496
2nd Ionisation Energy	4562
1st Electron Affinity	−81.1
2nd Electron Affinity	–
Enthalpy of Fusion, $\Delta H^{\ominus}_{Fusion}$	2.6
Enthalpy of Vaporisation, $\Delta H^{\ominus}_{Vaporisation}$	99.2
Enthalpy of Atomisation, $\Delta H^{\ominus}_{Atomisation}$	108.4
	J $K^{-1}mol^{-1}$
Standard Entropy, $S^{\ominus}$	51.2

Associated Bond Lengths (nm) & Energies (kJ mol^{-1})	n.a.

Strontium

[After Strontian, Scotland]

Atomic Number	38
Relative Atomic Mass	87.62

Chemical Symbol	Sr	Group	2	Period	5

Main Source & History

First isolated by Sir H. Davy in 1808, London, UK. It occurs in the ores strontianite ($SrCO_3$) and celestine ($SrSO_4$). The metal is obtained either by reduction of SrO with aluminium or by the electrolysis of the fused chloride.

Properties

Strontium is a soft, silvery–white metal. It is both malleable and ductile. Strontium is relatively reactive but less so than the alkali metals. It tarnishes quickly when exposed to the air, and reacts vigorously with water and acids from which it liberates hydrogen.

Uses

Strontium compounds are used in flares and fireworks ($Sr(NO_3)_2$ –to give a characteristic red colour, in glass and ceramics manufacture and sugar refining ($Sr(OH)_2$). Strontium alloys are used as permanent magnets.

Biological Role

Strontium has no known biological role. It should be noted that the danger of the radioisotope ^{90}Sr is that it can follow similar biological pathways to calcium.

Melting Point/K	1042.0	Boiling Point/K	1657.0	Density/g cm^{-3}	2.60

Electronic Configuration	[Kr] $5s^2$

Oxidation State(s)	0,±2	Isotopes	^{88}Sr(82.56%), ^{86}Sr(9.86%)

S.E.P.	$E^{\ominus}$/Volts
Sr^{2+}|Sr	− 2.89

Electronegativity (Pauling)	1.0

Radii	nm
Ionic	0.113 (+2)
Atomic	0.215
Van der Waals	–
Covalent	0.191

Enthalpy Data	kJ mol^{-1}
1st Ionisation Energy	550
2nd Ionisation Energy	1064
1st Electron Affinity	+146.0
2nd Electron Affinity	–
Enthalpy of Fusion, $\Delta H^{\ominus}_{Fusion}$	9.2
Enthalpy of Vaporisation, $\Delta H^{\ominus}_{Vaporisation}$	141.0
Enthalpy of Atomisation, $\Delta H^{\ominus}_{Atomisation}$	164.4
	J $K^{-1}mol^{-1}$
Standard Entropy, $S^{\ominus}$	52.3

Associated Bond Lengths (nm) & Energies (kJ mol^{-1})	n.a.

Sulphur/Sulfur

[Latin, 'sulphurium' = sulphur]

Atomic Number	16
Relative Atomic Mass	32.066

Chemical Symbol	S	Group	6	Period	3

Main Source & History

Sulphur has been known since pre–historic times. It occurs native as well as in the sulphide ores galena and zinc blende, and sulphate ores. Native sulphur is obtained by the Frasch Process. Superheated water at high pressure is forced into the sulphur deposit forcing out molten sulphur. Also extracted from H_2S found in natural gas by reaction with SO_2 over a catalyst.

Properties

Sulphur is a non–metallic element which occurs in several allotropic forms. Under normal conditions rhombic sulphur is the most stable form. It is a pale yellow, brittle, crystalline solid. Monoclinic sulphur is stable above the transition temperature of 369K and below 392K (its m.pt.). Sulphur is a reactive element and reacts with most other elements.

Uses

The chief use of sulphur is in the manufacture of sulphuric acid (Contact Process), which is used in the production of fertilisers, explosives, pigments, detergents, soaps, dyes and plastics. It is also used in the paper industry as well as in the manufacture of gunpowder, insecticides, matches, fungicides and vulcanised rubber.

Biological Role

Sulphur is found universally in organic compounds such as proteins in cells eg. keratin and many other organic compounds eg. co–enzyme A. It is found in animals and plants as the sulphate ion (SO_4^{2-}), the deficiency of which in plants causes chlorosis (yellowing of leaves).

Melting Point/K	386 rhombic 392 mon	Boiling Point/K	718.0	Density/g cm^{-3}	2.07 rhombic 1.96 mon

Electronic Configuration	[Ne] $3s^2 3p^4$

Oxidation State(s)	–2,0,+2,+3,±4,+5,±6	Isotopes	^{32}S(95.0%), ^{34}S(4.22%)

S.E.P.	$E^\ominus$/Volts
H_2SO_3|S	+0.51 Acid solution

Electronegativity (Pauling)	2.5

Radii	nm
Ionic	0.184 (–2)
Atomic	0.104
Van der Waals	0.180
Covalent	0.104

Enthalpy Data	kJ mol^{-1}
1st Ionisation Energy	1000
2nd Ionisation Energy	2251
1st Electron Affinity	–200.0
2nd Electron Affinity	+590.0
Enthalpy of Fusion, $\Delta H^\ominus_{Fusion}$	1.4
Enthalpy of Vaporisation, $\Delta H^\ominus_{Vaporisation}$	9.6
Enthalpy of Atomisation, $\Delta H^\ominus_{Atomisation}$	278.5
	J $K^{-1}mol^{-1}$
Standard Entropy, $S^\ominus$	32.6

Associated Bond Lengths (nm) & Energies (kJ mol^{-1})	S–S(0.208,226), S–H(0.134,347), S–O(0.143,265), S=O(0.144,525[2]), S–Cl(0.199,255), S–C(0.182,272)

Tantalum

[After mythical Greek king Tantalus]

Atomic Number	73
Relative Atomic Mass	180.9479

Chemical Symbol	Ta	Group	–	Period	6 3rd series T.M.

Main Source & History

Discovered by A.G. Ekeberg in 1802, Uppsala, Sweden. Occurs in the ore tantalite from which it is extracted by a multi–stage process involving sodium hydroxide fusion, acid washing and HF extraction. Finally the metal is obtained by alkali metal or carbon reduction of Ta_2O_5. It is also obtained as a by–product of tin extraction from cassiterite.

Properties

Tantalum is a lustrous, grey–white metal which is very malleable and ductile. The metal is relatively unreactive but will react with a mixture of nitric and hydrofluoric acids.

Uses

Tantalum is mainly used in the production of corrosion resistant alloys which are used, for example, in the construction of chemical plants. The alloys are also used in the manufacture of surgical and dental instruments. As a carbide tantalum is used in conjunction with tungsten carbide for producing steel cutting tools.

Biological Role

Tantalum has no known biological role.

Melting Point/K	3269.0	Boiling Point/K	5700.0^2	Density/g cm^{-3}	16.65^{293K}

Electronic Configuration	[Xe] $4f^{14}$, $5d^3$, $6s^2$

Oxidation State(s)	0, +1, +3, +4, <u>+5</u>	Isotopes	^{181}Ta(99.99%), ^{180}Ta(trace)

S.E.P.	$E^{\ominus}$/Volts
Ta_2O_5|Ta	– 0.81

Electronegativity (Pauling)	1.5

Radii	nm
Ionic	0.068 (+5)
Atomic	0.143
Van der Waals	–
Covalent	0.134

Enthalpy Data	kJ mol^{-1}
1st Ionisation Energy	761
2nd Ionisation Energy	1560
1st Electron Affinity	+60.0
2nd Electron Affinity	–
Enthalpy of Fusion, $\Delta H^{\ominus}_{Fusion}$	31.4
Enthalpy of Vaporisation, $\Delta H^{\ominus}_{Vaporisation}$	753.1
Enthalpy of Atomisation, $\Delta H^{\ominus}_{Atomisation}$	782.0
	J K^{-1} mol^{-1}
Standard Entropy, $S^{\ominus}$	41.5

Associated Bond Lengths (nm) & Energies (kJ mol^{-1})	n.a.

Technetium

[Greek, 'technikos' = artificial]

Atomic Number	43
Relative Atomic Mass	98.9063[1]

Chemical Symbol	Tc	Group	–	Period	5 (2nd series T.M.)

Main Source & History

Discovered by C. Perrier and E. Segré in 1937, Italy. Technetium is obtained in large quantities from the fission products of uranium in nuclear reactors. The metal is obtained by reduction of some of its compounds with hydrogen.

Properties

Technetium is a silvery–grey, radioactive metal. The most stable isotope ^{97}Tc has a $T_{1/2}$ of 2.6×10^6 years. It dissolves in hydrogen peroxide and is attacked slowly by moist air. Technetium is reactive towards oxidising acids.

Uses

The man made isotope ^{99}Tc is used in clinical medicine as both a tracer and radiation source. Pertechnetates containing the TcO_4^- ion are used in alloys as corrosion inhibitors.

Biological Role

Technetium has no known biological role. It is hazardous to health due to its radioactive nature.

Melting Point/K	2445.0	Boiling Point/K	5150.0	Density/g cm^{-3}	11.50[4]

Electronic Configuration	[Kr] $4d^6, 5s^1$

Oxidation State(s)	– 1,0,±4,±5,+6,±7	Isotopes	^{97}Tc(0%), ^{98}Tc(0%)

S.E.P.	$E^\ominus$/Volts
TcO_4^- | Tc	+ 0.47

Electronegativity (Pauling)	1.9

Radii	nm
Ionic	0.056 (+7)
Atomic	0.135
Van der Waals	–
Covalent	0.127

Enthalpy Data	kJ mol^{-1}
1st Ionisation Energy	702
2nd Ionisation Energy	1472
1st Electron Affinity	+70.0
2nd Electron Affinity	–
Enthalpy of Fusion, $\Delta H^\ominus_{Fusion}$	23.8
Enthalpy of Vaporisation, $\Delta H^\ominus_{Vaporisation}$	577.4
Enthalpy of Atomisation, $\Delta H^\ominus_{Atomisation}$	678.0
	J $K^{-1}mol^{-1}$
Standard Entropy, $S^\ominus$	33.5

Associated Bond Lengths (nm) & Energies (kJ mol^{-1})	n.a.

Tellurium

[Latin, 'tellus' = earth]

Atomic Number	52
Relative Atomic Mass	127.60

Chemical Symbol	Te	Group	6	Period	5

Main Source & History

Discovered by Baron von Reichenstein in 1783, Romania. Tellurium occurs as tellurides such as tetradymite (Bi_2Te_3) as well as in sulphide ores. It is extracted from anode mud and flue dusts from copper refining using sulphuric acid. The element is obtained by reduction with zinc.

Properties

Tellurium is a silvery–white, brittle metalloid. The metallic form is grey in colour and has low electrical conductivity. Tellurium reacts readily with halogens, oxygen and some metals and is attacked by oxidising acids.

Uses

Tellurium is used in the manufacture of alloys particularly lead and copper where it improves their tensile strength.

Biological Role

Tellurium has no known biological role. It is a very toxic element whose compounds are thought to cause malformation of the foetus.

Melting Point/K	723.0	Boiling Point/K	1263.0	Density/g cm^{-3}	6.25

Electronic Configuration	[Kr] $4d^{10}$, $5s^2 5p^4$

Oxidation State(s)	−2, −1, 0, +2, <u>±4</u>, +5, +6	Isotopes	^{130}Te(34.49%), ^{128}Te(31.69%)

S.E.P.	$E^{\ominus}$/Volts
Te^{4+}|Te	+ 0.57 (Acid solution)

Electronegativity (Pauling)	2.1

Radii	nm
Ionic	0.097 (+4)
Atomic	0.143
Van der Waals	0.220
Covalent	0.137

Enthalpy Data	kJ mol^{-1}
1st Ionisation Energy	869
2nd Ionisation Energy	1795
1st Electron Affinity	−190.2
2nd Electron Affinity	–
Enthalpy of Fusion, $\Delta H^{\ominus}_{Fusion}$	17.5
Enthalpy of Vaporisation, $\Delta H^{\ominus}_{Vaporisation}$	49.7
Enthalpy of Atomisation, $\Delta H^{\ominus}_{Atomisation}$	196.7
	J $K^{-1}mol^{-1}$
Standard Entropy, $S^{\ominus}$	49.7

Associated Bond Lengths (nm) & Energies (kJ mol^{-1})	Te–Te(0.286,235), Te–H(0.170,240[2]), Te–O(0.200,268), Te–Cl(0.231,251)

Terbium

[After Ytterby in Sweden]

Atomic Number	65
Relative Atomic Mass	158.9253

Chemical Symbol	Tb	Group	–	Period	6 Lanthanide series

Main Source & History

Discovered by C.G. Mosander in 1843, Stockholm, Sweden. The main ores are monazite and bastnaesite from which terbium is extracted, with difficulty, by ion exchange.

Properties

Terbium is a silvery–white metal which is a good conductor of heat and electricity. Terbium reacts slowly with oxygen and water liberating hydrogen.

Uses

Terbium compounds have uses as phosphors (Tb_2O_3) in colour televisions and as a lasing agents.

Biological Role

Terbium has no known biological role.

Melting Point/K	1629.0	Boiling Point/K	3100.0[2]	Density/g cm^{-3}	8.27

Electronic Configuration	[Xe] $4f^9$, $6s^2$

Oxidation State(s)	0, <u>±3</u>, +4	Isotopes	^{159}Tb(100%)

S.E.P.	$E^{\ominus}$/Volts
Tb^{3+}|Tb	– 2.31 (Acid solution)

Electronegativity (Pauling)	1.2

Radii	nm
Ionic	0.093 (+3)
Atomic	0.177
Van der Waals	–
Covalent	0.159

Enthalpy Data	kJ mol^{-1}
1st Ionisation Energy	564
2nd Ionisation Energy	1112
1st Electron Affinity	+50.0[2]
2nd Electron Affinity	–
Enthalpy of Fusion, $\Delta H^{\ominus}_{Fusion}$	16.3
Enthalpy of Vaporisation, $\Delta H^{\ominus}_{Vaporisation}$	391.0
Enthalpy of Atomisation, $\Delta H^{\ominus}_{Atomisation}$	–
	J $K^{-1}mol^{-1}$
Standard Entropy, $S^{\ominus}$	73.2

Associated Bond Lengths (nm) & Energies (kJ mol^{-1})	n.a.

Thallium

[Greek, 'thallos' = a green twig]

Atomic Number	81
Relative Atomic Mass	204.3833

Chemical Symbol	Tl	Group	3	Period	6

Main Source & History

Discovered by W. Crookes in 1861, London, UK. Thallium occurs in some sulphide and selenide ores eg. crookesite. It is extracted from flue dusts produced by the roasting of certain sulphide ores (galena and zinc blende). The metal is obtained by electrolysis.

Properties

Thallium is a soft, white, malleable metal resembling lead. Thallium is a relatively reactive metal. It is attacked readily by moist air and acids especially those of an oxidising nature.

Uses

Thallium is not used due to its extreme toxicity. Compounds of thallium are used as rat poisons (TlCl) and insecticides, as well as in the manufacture of special glasses.

Biological Role

Thallium has no known biological role. It is an extremely toxic element.

Melting Point/K	576.7	Boiling Point/K	1730.0	Density/g cm^{-3}	11.86

Electronic Configuration	[Xe] $4f^{14}$, $5d^{10}$, $6s^2 6p^1$

Oxidation State(s)	0,±1,+3	Isotopes	^{205}Tl(70.48%),^{203}Tl(29.52%)

S.E.P.	$E^{\ominus}$/Volts
Tl^+|Tl	− 0.34 (Acid solution)

Electronegativity (Pauling)	1.8

Radii	nm
Ionic	0.149 (+1)
Atomic	0.170
Van der Waals	–
Covalent	0.155

Enthalpy Data	kJ mol^{-1}
1st Ionisation Energy	589
2nd Ionisation Energy	1971
1st Electron Affinity	−30.0
2nd Electron Affinity	–
Enthalpy of Fusion, $\Delta H^{\ominus}_{Fusion}$	4.3
Enthalpy of Vaporisation, $\Delta H^{\ominus}_{Vaporisation}$	162.1
Enthalpy of Atomisation, $\Delta H^{\ominus}_{Atomisation}$	182.2
	J K^{-1} mol^{-1}
Standard Entropy, $S^{\ominus}$	64.2

Associated Bond Lengths (nm) & Energies (kJ mol^{-1})	Tl–Tl(0.341,63^2), Tl– H(0.187,185)

Thorium

[After the Norse god of thunder, Thor]

Atomic Number	90
Relative Atomic Mass	232.0381[1]

Chemical Symbol	Th	Group	–	Period	7 Actinide series

Main Source & History

Discovered by J.J. Berzelius in 1818, Stockholm, Sweden. The principal minerals are thorite ($ThSiO_4$) and monazite. It is extracted by a multi-stage process involving dissolving the ores in concentrated H_2SO_4 and extraction using tributylphosphate. ThO_2 is obtained by heating the phosphate with Na_2CO_3. The metal is obtained by reduction of $ThCl_4$ with sodium.

Properties

Thorium is a soft, silvery, radioactive metal. It is both malleable and ductile. The most stable isotope, ^{232}Th has a $T_{1/2}$ of 1.4×10^{10} years. This unreactive element is slowly attacked by air as well as steam from which hydrogen is liberated. It also reacts with acids.

Uses	Thorium is used as an oxygen 'getter' in the electronics industry. ThO_2 is used in the manufacture of alloys, refractories and special glasses. It is also used as a catalyst for the manufacture of hydrocarbons from coal, lignite or natural gas (Fischer–Tropsch Process).
Biological Role	Thorium has no known biological role. It is hazardous to health due to its radioactive nature.

Melting Point/K	2023.0	Boiling Point/K	5060.0[2]	Density/g cm^{-3}	11.70

Electronic Configuration	[Rn] $6d^2$, $7s^2$

Oxidation State(s)	0,±4	Isotopes	^{232}Th(99.99%), ^{230}Th(trace)

S.E.P.	$E^\ominus$/Volts
Th^{4+}|Th	– 1.83 Acid solution

Electronegativity (Pauling)	1.3

Radii	nm
Ionic	0.099(+4)
Atomic	0.180
Van der Waals	–
Covalent	0.165

Enthalpy Data	kJ mol^{-1}
1st Ionisation Energy	671
2nd Ionisation Energy	1110
1st Electron Affinity	–
2nd Electron Affinity	–
Enthalpy of Fusion, $\Delta H^\ominus_{Fusion}$	15.7[2]
Enthalpy of Vaporisation, $\Delta H^\ominus_{Vaporisation}$	543.9
Enthalpy of Atomisation, $\Delta H^\ominus_{Atomisation}$	598.3
	J $K^{-1}mol^{-1}$
Standard Entropy, $S^\ominus$	53.4

Associated Bond Lengths (nm) & Energies (kJ mol^{-1})	n.a.

Thulium

[Latin, 'Thule' = Northland]

Atomic Number	69
Relative Atomic Mass	168.9342

Chemical Symbol	Tm	Group	–	Period	6 Lanthanide series

Main Source & History

Discovered by P.T. Cleve in 1880, Uppsala, Sweden. Thulium is the rarest member of the lanthanide series. The main ores are monazite and bastnaesite from which thulium is extracted, with extreme difficulty, by ion exchange.

Properties

Thulium is a silvery coloured metal which is a relatively good conductor of heat and electricity. Thulium reacts with both air and water from which hydrogen is liberated.

Uses

After neutron irradiation thulium and its compounds (such as Tm_2O_3) are used as portable x–ray sources.

Biological Role

Thulium has no known biological role.

Melting Point/K	1818.0	Boiling Point/K	2000.0[2]	Density/g cm^{-3}	9.30

Electronic Configuration	[Xe] $4f^{13}$, $6s^2$

Oxidation State(s)	0, +2, <u>+3</u>	Isotopes	^{169}Tm(100%)

S.E.P.	$E^{\ominus}$/Volts
Tm^{3+}|Tm	– 2.31 Acid solution

Electronegativity (Pauling)	1.2

Radii	nm
Ionic	0.087(+3)
Atomic	0.174
Van der Waals	–
Covalent	0.156

Enthalpy Data	kJ mol^{-1}
1st Ionisation Energy	596
2nd Ionisation Energy	1163
1st Electron Affinity	+50.0[2]
2nd Electron Affinity	–
Enthalpy of Fusion, $\Delta H^{\ominus}_{Fusion}$	18.4
Enthalpy of Vaporisation, $\Delta H^{\ominus}_{Vaporisation}$	247.0
Enthalpy of Atomisation, $\Delta H^{\ominus}_{Atomisation}$	–
	J $K^{-1}mol^{-1}$
Standard Entropy, $S^{\ominus}$	74.0

Associated Bond Lengths (nm) & Energies (kJ mol^{-1})	n.a.

Tin

[Latin, 'stannum' = tin (Anglo–Saxon)]

Atomic Number	50
Relative Atomic Mass	118.710

Chemical Symbol	Sn	Group	4	Period	5

Main Source & History

Tin has been known since pre–historic times. The main ore is cassiterite (SnO_2) from which the metal is obtained by reduction with carbon. Tin is refined by applying just sufficient heat to melt it in a reverberatory furnace or electrolytically.

Properties

Tin is a silvery–white metal with a pale blue tinge. It is soft, malleable and ductile. Another physical form, known as grey tin, which is a brittle grey powder,exists below 291K. Tin is unreactive to oxygen and water, but reacts readily with concentrated acids. It reacts with hot concentrated alkalis liberating hydrogen.

Uses

Tin is used in alloying (eg. solder, Type Metal, bronze), with niobium it acts as a superconductor, and it is also used for tin plating (food containers). Tin compounds are used in fungicides (a variety of organic derivatives), as polymer additives, in special glasses, in dyeing ($SnCl_2$), in toothpaste (SnF_2) and in pigments for paints (SnS_2).

Biological Role

Tin is suggested to be essential to humans but its exact biological role is unclear. High concentrations of tin and its compounds are known to be toxic and carcinogenic. The recommended limit is 250 ppm in canned foods.

Melting Point/K	505.0	Boiling Point/K	2543.0	Density/g cm^{-3}	7.28

Electronic Configuration	[Kr] $4d^{10}$, $5s^2 5p^2$

Oxidation State(s)	− 4,0,±2,±4	Isotopes	^{120}Sn(32.4%),^{118}Sn(24.3%)

S.E.P.	$E^{\ominus}$/Volts
SnO|Sn	− 0.14 (Acid solution)

Electronegativity (Pauling)	1.8

Radii	nm
Ionic	0.093(+2)
Atomic	0.140
Van der Waals	–
Covalent	0.140

Enthalpy Data	kJ mol^{-1}
1st Ionisation Energy	709
2nd Ionisation Energy	1412
1st Electron Affinity	−121.0
2nd Electron Affinity	–
Enthalpy of Fusion, $\Delta H^{\ominus}_{Fusion}$	7.2
Enthalpy of Vaporisation, $\Delta H^{\ominus}_{Vaporisation}$	290.4
Enthalpy of Atomisation, $\Delta H^{\ominus}_{Atomisation}$	302.0
	J $K^{-1}mol^{-1}$
Standard Entropy, $S^{\ominus}$	51.5

Associated Bond Lengths (nm) & Energies (kJ mol^{-1})	Sn–Sn(0.281,195), Sn– H(0.170,314[2])

Titanium

[Latin, 'titanes' = sons of the earth]

Atomic Number	22
Relative Atomic Mass	47.88

Chemical Symbol	Ti	Group	–	Period	4 1st series T.M.

Main Source & History

Discovered by W. Gregor in 1791, Cornwall, UK. The main ores are rutile (TiO_2) and ilmenite ($FeTiO_3$). TiO_2 is converted to the chloride which is reduced with magnesium, calcium or sodium to the metal (Kroll Process).

Properties

Titanium is a hard, silvery, lustrous metal of low density. It is corrosion resistant although the fine powdered metal will burn readily in oxygen. It will liberate hydrogen from steam at high temperature. It is not readily attacked by acids although it will dissolve in concentrated hydrochloric and sulphuric acids.

Uses

Titanium is used in the manufacture of strong, light alloys for use in aircraft and missile manufacture as well as for chemical plants and car engines. Compounds of titanium are used as pigments in paints (TiO_2) and as a food additive (TiO_2). This oxide is also used in the paper, ceramics and textile industry. Other compounds are used in the polymerisation of alkenes ($TiCl_4$ – Ziegler Process) and for the manufacture of high speed tools (TiC).

Biological Role

Titanium has no known biological role although it is suggested that its compounds may be carcinogenic.

Melting Point/K	1933.0	Boiling Point/K	3560.0	Density/g cm^{-3}	4.54

Electronic Configuration	[Ar] $3d^2$, $4s^2$

Oxidation State(s)	0, +2, +3, <u>+4</u>	Isotopes	^{48}Ti(73.99%), ^{46}Ti(8.2%)

S.E.P.	$E^{\ominus}$/Volts
TiO^{2+}|Ti	– 0.86 Acid solution

Electronegativity (Pauling)	1.5

Radii	nm
Ionic	0.068(+4)
Atomic	0.145
Van der Waals	–
Covalent	0.132

Enthalpy Data	kJ mol^{-1}
1st Ionisation Energy	658
2nd Ionisation Energy	1310
1st Electron Affinity	+20.0
2nd Electron Affinity	–
Enthalpy of Fusion, $\Delta H^{\ominus}_{Fusion}$	15.0
Enthalpy of Vaporisation, $\Delta H^{\ominus}_{Vaporisation}$	427.0
Enthalpy of Atomisation, $\Delta H^{\ominus}_{Atomisation}$	469.9
	J $K^{-1}mol^{-1}$
Standard Entropy, $S^{\ominus}$	30.6

Associated Bond Lengths (nm) & Energies (kJ mol^{-1})	n.a.

Tungsten/Wolfram

[Swedish, 'tungsten' = heavy stone]

Atomic Number	74
Relative Atomic Mass	183.85

Chemical Symbol	W	Group	–	Period	6 3rd series T.M.

Main Source & History

Discovered by J.J. and F. d'Elhuyar in 1783, Spain. The main ores of tungsten are wolframite ($(Fe,Mn)WO_4$) and scheelite ($CaWO_4$). WO_3 is precipitated with acid after the concentrated ores are fused with sodium hydroxide. The metal is obtained by the reduction of WO_3 with hydrogen.

Properties

Tungsten is a silvery–white, lustrous metal. It is hard, malleable and ductile and has the highest melting point of any metal. It is very resistant to corrosion, even to attack by acids.

Uses

Tungsten is used in the manufacture of alloys which are used for high speed cutting tools. It is used as a filament in electric light bulbs (also H_2WO_4) and as electrical contacts. Compounds of tungsten are used in the manufacture of high speed cutting tools (W_2C), for fireproofing fabrics, in glazes (WO_3) and as mordants in dyeing.

Biological Role

Tungsten has no known biological role.

Melting Point/K	3680.0^2	Boiling Point/K	6200.0	Density/g cm^{-3}	19.40

Electronic Configuration	[Xe] $4f^{14}$, $5d^4$, $6s^2$

Oxidation State(s)	–2, –1, 0, +2, +3, +5, $\underline{+6}$	Isotopes	^{184}W(30.7%), ^{186}W(28.6%)

S.E.P.	$E^\ominus$/Volts
WO_3|W	–0.09 Acid solution

Electronegativity (Pauling)	1.7

Radii	nm
Ionic	0.062(+6)
Atomic	0.137
Van der Waals	–
Covalent	0.130

Enthalpy Data	kJ mol^{-1}
1st Ionisation Energy	770
2nd Ionisation Energy	1700
1st Electron Affinity	+60.0
2nd Electron Affinity	–
Enthalpy of Fusion, $\Delta H^\ominus_{Fusion}$	35.2
Enthalpy of Vaporisation, $\Delta H^\ominus_{Vaporisation}$	799.1
Enthalpy of Atomisation, $\Delta H^\ominus_{Atomisation}$	849.4
	J $K^{-1}mol^{-1}$
Standard Entropy, $S^\ominus$	32.6

Associated Bond Lengths (nm) & Energies (kJ mol^{-1})	n.a.

Unnilennium

Atomic Number	109
Relative Atomic Mass	–

Chemical Symbol	Une	Group	–	Period	7 4th series T.M.

Main Source & History

Artificially produced in a nuclear particle accelerator in a similar way to ^{261}Unq.

Properties

Unnilennium is a radioactive metal. The most stable isotope, so far produced, ^{266}Une has a $T_{1/2}$ of 5 milliseconds.

Uses	Unnilennium has no known uses.

Biological Role	Unnilennium has no known biological role. It is hazardous to health due to its extreme radioactive nature.

Melting Point/K	–	Boiling Point/K	–	Density/g cm^{-3}	–

Electronic Configuration	–

Oxidation State(s)	–	Isotopes	–

S.E.P. $E^{\ominus}$/Volts
n.a.

Electronegativity (Pauling)	–

Radii	nm
Ionic	–
Atomic	–
Van der Waals	–
Covalent	–

Enthalpy Data	kJ mol^{-1}
1st Ionisation Energy	–
2nd Ionisation Energy	–
1st Electron Affinity	–
2nd Electron Affinity	–
Enthalpy of Fusion, $\Delta H^{\ominus}_{Fusion}$	–
Enthalpy of Vaporisation, $\Delta H^{\ominus}_{Vaporisation}$	–
Enthalpy of Atomisation, $\Delta H^{\ominus}_{Atomisation}$	–
	J $K^{-1}mol^{-1}$
Standard Entropy, $S^{\ominus}$	–

Associated Bond Lengths (nm) & Energies (kJ mol^{-1})	n.a.

Unnilhexium

Atomic Number	106
Relative Atomic Mass	263.1182[1]

Chemical Symbol	Unh	Group	–	Period	7 4th series T.M.

Main Source & History

Discovered by G.N. Flerov (USSR) and A. Ghiorso (USA) in 1974. Artificially produced in a nuclear particle accelerator in a similar way to ^{261}Unq.

Properties

Unnilhexium is a radioactive metal. The most stable isotope, so far produced, ^{263}Unh has a $T_{1/2}$ of 1 second.

Uses

Unnilhexium has no known uses.

Biological Role

Unnilhexium has no known biological role. It is hazardous to health due to its extreme radioactive nature.

Melting Point/K	–	Boiling Point/K	–	Density/g cm^{-3}	–

Electronic Configuration	[Rn] $5f^{14}$, $6d^{4}$, $7s^{2}$

Oxidation State(s)	–	Isotopes	–

S.E.P. $E^{\ominus}$/Volts
n.a.

Electronegativity (Pauling)	–

Radii	nm
Ionic	–
Atomic	–
Van der Waals	–
Covalent	–

Enthalpy Data	kJ mol^{-1}
1st Ionisation Energy	–
2nd Ionisation Energy	–
1st Electron Affinity	–
2nd Electron Affinity	–
Enthalpy of Fusion, $\Delta H^{\ominus}_{Fusion}$	–
Enthalpy of Vaporisation, $\Delta H^{\ominus}_{Vaporisation}$	–
Enthalpy of Atomisation, $\Delta H^{\ominus}_{Atomisation}$	–
	J $K^{-1}mol^{-1}$
Standard Entropy, $S^{\ominus}$	–

Associated Bond Lengths (nm) & Energies (kJ mol^{-1})	n.a.

Unniloctium

Atomic Number	108
Relative Atomic Mass	–

Chemical Symbol	Uno	Group	–	Period	7 4th series T.M.

Main Source & History

Artificially produced in a nuclear particle accelerator in a similar way to ^{261}Unq.

Properties

Unniloctium is a radioactive metal. The most stable isotope, so far produced, ^{265}Uno has a $T_{1/2}$ of 2 milliseconds.

Uses

Unniloctium has no known uses.

Biological Role

Unniloctium has no known biological role. It is hazardous to health due to its extreme radioactive nature.

Melting Point/K	–	Boiling Point/K	–	Density/g cm^{-3}	–

Electronic Configuration	–

Oxidation State(s)	–	Isotopes	–

S.E.P. $E^{\ominus}$/Volts
n.a.

Electronegativity (Pauling)	–

Radii	nm
Ionic	–
Atomic	–
Van der Waals	–
Covalent	–

Enthalpy Data	kJ mol^{-1}
1st Ionisation Energy	–
2nd Ionisation Energy	–
1st Electron Affinity	–
2nd Electron Affinity	–
Enthalpy of Fusion, $\Delta H^{\ominus}_{Fusion}$	–
Enthalpy of Vaporisation, $\Delta H^{\ominus}_{Vaporisation}$	–
Enthalpy of Atomisation, $\Delta H^{\ominus}_{Atomisation}$	–
	J K^{-1}mol^{-1}
Standard Entropy, $S^{\ominus}$	–

Associated Bond Lengths (nm) & Energies (kJ mol^{-1})	n.a.

Unnilpentium

(Hahnium/Nielsbohrium)

Atomic Number	105
Relative Atomic Mass	262.1138[1]

Chemical Symbol	Unp (Ha/Ns)	Group	–	Period	7 4th series T.M.

Main Source & History

Discovered by G.N. Flerov et al. (USSR) in 1970 and A. Ghiorso et al. (USA) in 1970. Artificially produced in a nuclear particle accelerator in a similar way to ^{261}Unq.

Properties

Unnilpentium is a radioactive metal. The most stable isotope, so far produced, ^{262}Unp has a $T_{1/2}$ of 34 seconds.

Uses

Unnilpentium has no known uses.

Biological Role

Unnilpentium has no known biological role. It is hazardous to health due to its extreme radioactive nature.

Melting Point/K	–	Boiling Point/K	–	Density/g cm^{-3}	–

Electronic Configuration	[Rn] $5f^{14}$, $6d^{3}$, $7s^{2}$

Oxidation State(s)	–	Isotopes	–

S.E.P. $E^{\ominus}$/Volts
n.a.

Electronegativity (Pauling)	–

Radii	nm
Ionic	–
Atomic	–
Van der Waals	–
Covalent	–

Enthalpy Data	kJ mol^{-1}
1st Ionisation Energy	–
2nd Ionisation Energy	–
1st Electron Affinity	–
2nd Electron Affinity	–
Enthalpy of Fusion, $\Delta H^{\ominus}_{Fusion}$	–
Enthalpy of Vaporisation, $\Delta H^{\ominus}_{Vaporisation}$	–
Enthalpy of Atomisation, $\Delta H^{\ominus}_{Atomisation}$	–
	J $K^{-1}mol^{-1}$
Standard Entropy, $S^{\ominus}$	–

Associated Bond Lengths (nm) & Energies (kJ mol^{-1})	n.a.

Unnilquadium

(Rutherfordium/Kurchatovium)

Atomic Number	104
Relative Atomic Mass	261.1087[1]

Chemical Symbol	Unq (Rf/Ku)	Group	–	Period	7 4th series T.M.

Main Source & History

Discovered by G.N. Flerov et al. (USSR) in 1964 and A. Ghiorso et al. (USA) in 1969. Artificially produced in a nuclear particle accelerator from bombardment of ^{242}Pa with ^{22}Ne.

Properties

Unnilquadium is a radioactive metal. The most stable isotope, so far produced, ^{261}Unq has a $T_{1/2}$ of 65 seconds.

Uses	Unnilquadium has no known uses.

Biological Role	Unnilquadium has no known biological role. It is hazardous to health due to its extreme radioactive nature.

Melting Point/K	–	Boiling Point/K	–	Density/g cm^{-3}	–

Electronic Configuration	[Rn] $5f^{14}$, $6d^2$, $7s^2$

Oxidation State(s)	–	Isotopes	–

S.E.P. $E^{\ominus}$/Volts
n.a.

Electronegativity (Pauling)	–

Radii	nm
Ionic	–
Atomic	–
Van der Waals	–
Covalent	–

Enthalpy Data	kJ mol^{-1}
1st Ionisation Energy	–
2nd Ionisation Energy	–
1st Electron Affinity	–
2nd Electron Affinity	–
Enthalpy of Fusion, $\Delta H^{\ominus}_{Fusion}$	–
Enthalpy of Vaporisation, $\Delta H^{\ominus}_{Vaporisation}$	–
Enthalpy of Atomisation, $\Delta H^{\ominus}_{Atomisation}$	–
	J $K^{-1}mol^{-1}$
Standard Entropy, $S^{\ominus}$	–

Associated Bond Lengths (nm) & Energies (kJ mol^{-1})	n.a.

Unnilseptium

Atomic Number	107
Relative Atomic Mass	262.1229[1]

Chemical Symbol	Uns	Group	–	Period	7 4th series T.M.

Main Source & History

Artificially produced in a nuclear particle accelerator in a similar way to ^{261}Unq.

Properties

Unnilseptium is a radioactive metal. The most stable isotope, so far produced, ^{262}Unq has a $T_{1/2}$ of 115 milliseconds.

Uses	Unnilseptium has no known uses.

Biological Role	Unnilseptium has no known biological role. It is hazardous to health due to its extreme radioactive nature.

Melting Point/K	–	Boiling Point/K	–	Density/g cm^{-3}	–

Electronic Configuration	–

Oxidation State(s)	–	Isotopes	–

S.E.P. $E^{\ominus}$/Volts
n.a.

Electronegativity (Pauling)	–

Radii	nm
Ionic	–
Atomic	–
Van der Waals	–
Covalent	–

Enthalpy Data	kJ mol^{-1}
1st Ionisation Energy	–
2nd Ionisation Energy	–
1st Electron Affinity	–
2nd Electron Affinity	–
Enthalpy of Fusion, $\Delta H^{\ominus}_{Fusion}$	–
Enthalpy of Vaporisation, $\Delta H^{\ominus}_{Vaporisation}$	–
Enthalpy of Atomisation, $\Delta H^{\ominus}_{Atomisation}$	–
	J $K^{-1} mol^{-1}$
Standard Entropy, $S^{\ominus}$	–

Associated Bond Lengths (nm) & Energies (kJ mol^{-1})	n.a.

Uranium

[After the planet Uranus]

Atomic Number	92
Relative Atomic Mass	238.0289[1]

Chemical Symbol	U	Group	–	Period	7 Actinide series

Main Source & History

Discovered by E–M. Peligot in 1842, Paris, France. The important ores are pitch blende (U_3O_8) and uraninite (UO_2). The purified oxide is converted to UF_4 from which the metal is obtained by reduction with magnesium.

Properties

Uranium is a hard, silvery–white, radioactive metal which is both malleable and ductile. The most stable isotope ^{238}U has a $T_{1/2}$ of 4.5×10^9 years. It is relatively reactive metal. Uranium is readily oxidised as well as being attacked by acids.

Uses

Both ^{235}U and ^{238}U are important in the nuclear power and weapons industry. Compounds of uranium are used in special glasses ($Na_2U_2O_7.6H_2O$) and as catalysts (uranium carbide can be used in this manner in the synthesis of ammonia).

Biological Role

Uranium has no known biological role. It is hazardous to health due to its radioactive nature.

Melting Point/K	1405.3	Boiling Point/K	4091.0	Density/g cm^{-3}	19.05

Electronic Configuration	[Rn] $5f^3$, $6d^1$, $7s^2$

Oxidation State(s)	0, +3, $\underline{+4}$, +5, $\underline{+6}$	Isotopes	^{238}U(99.28%), ^{235}U(0.72%)

S.E.P.	$E^{\ominus}$/Volts
U^{4+}|U	– 1.38 (Acid solution)

Electronegativity (Pauling)	1.7

Radii	nm
Ionic	0.097(+4)
Atomic	0.139
Van der Waals	–
Covalent	0.142

Enthalpy Data	kJ mol^{-1}
1st Ionisation Energy	586
2nd Ionisation Energy	1420
1st Electron Affinity	+90.7
2nd Electron Affinity	–
Enthalpy of Fusion, $\Delta H^{\ominus}_{Fusion}$	15.5
Enthalpy of Vaporisation, $\Delta H^{\ominus}_{Vaporisation}$	417.1
Enthalpy of Atomisation, $\Delta H^{\ominus}_{Atomisation}$	535.6
	J $K^{-1} mol^{-1}$
Standard Entropy, $S^{\ominus}$	50.2

Associated Bond Lengths (nm) & Energies (kJ mol^{-1})	n.a.

Vanadium

[After the Norse goddess Vanadis]

Atomic Number	23
Relative Atomic Mass	50.9415

Chemical Symbol	V	Group	–	Period	4 (1st series T.M.)

Main Source & History	Properties
Discovered by A,M, del Rio in 1801, Mexico. The principal ores are carnotite ($KUO_2VO_4.1.5H_2O$), patronite (VS_4) and vanadinite ($Pb_5(VO_4)_3Cl$). Following a multi–stage process vanadium(V) oxide is reduced with calcium (in calcium chloride flux) to the metal.	Vanadium is a silvery–grey metal. It is both malleable and ductile. It is resistant to corrosion but dissolves in hot acids both dilute and concentrated.

Uses	Vanadium has extensive uses in the manufacture of steel alloys for the construction of exhaust valves and high speed tools. Compounds of vanadium are used as catalysts (for example, vanadium(V) oxide in the Contact Process for the manufacture of sulphuric acid) and in red phosphors for colour televisions (YVO_4:Eu).

Biological Role	Vanadium is suggested to be essential to humans but its exact biological role is uncertain. It is found in pigments of lower life forms.

Melting Point/K	2160.0^2	Boiling Point/K	3650.0^2	Density/g cm^{-3}	5.96

Electronic Configuration	[Ar] $3d^3$, $4s^2$

Oxidation State(s)	– 1,0,+2,+3,+4,$\underline{+5}$	Isotopes	^{51}V(99.75%),^{50}V(0.25%)

S.E.P.	$E^\ominus$/Volts
VO_2^+|V	– 0.24 (Acid solution)

Electronegativity (Pauling)	1.6

Radii	nm
Ionic	0.059(+5)
Atomic	0.131
Van der Waals	–
Covalent	0.122

Enthalpy Data	kJ mol^{-1}
1st Ionisation Energy	650
2nd Ionisation Energy	1414
1st Electron Affinity	+50.0
2nd Electron Affinity	–
Enthalpy of Fusion, $\Delta H^\ominus_{Fusion}$	17.6
Enthalpy of Vaporisation, $\Delta H^\ominus_{Vaporisation}$	458.6
Enthalpy of Atomisation, $\Delta H^\ominus_{Atomisation}$	514.2
	J $K^{-1}mol^{-1}$
Standard Entropy, $S^\ominus$	28.9

Associated Bond Lengths (nm) & Energies (kJ mol^{-1})	n.a.

Xenon

[Greek, 'xenos' = stranger]

Atomic Number	54
Relative Atomic Mass	131.29

Chemical Symbol	Xe	Group	0 (8)	Period	5

Main Source & History

Discovered by Sir W. Ramsay and M.W. Travers in 1898, London, UK. Obtained by the fractional distillation of liquid air.

Properties

Xenon is a colourless, odourless, tasteless gas. It is denser than air and moderately soluble in water. It is the most reactive of the naturally occuring noble gases, forming compounds with fluorine (XeF_4) and oxygen (XeO_3).

Uses

Xenon is used in the manufacture of discharge tubes and stroboscopic lamps.

Biological Role

Xenon has no known biological role, it is, however, known to be non– toxic.

Melting Point/K	161.0	Boiling Point/K	166.0	Density/g cm^{-3}	2.94^{166K}

Electronic Configuration	[Kr] $4d^{10}$, $5s^2 5p^6$

Oxidation State(s)	0, +2, ±4, ±6, +8	Isotopes	^{132}Xe(26.9%), ^{129}Xe(26.4%)

S.E.P.	$E^{\ominus}$/Volts
XeO_3|Xe	+2.12 (Acid solution)

Electronegativity (Pauling)	–

Radii	nm
Ionic	–
Atomic	0.221
Van der Waals	0.220
Covalent	0.130

Enthalpy Data	kJ mol^{-1}
1st Ionisation Energy	1170
2nd Ionisation Energy	2047
1st Electron Affinity	$+41.0^4$
2nd Electron Affinity	–
Enthalpy of Fusion, $\Delta H^{\ominus}_{Fusion}$	2.3
Enthalpy of Vaporisation, $\Delta H^{\ominus}_{Vaporisation}$	12.7
Enthalpy of Atomisation, $\Delta H^{\ominus}_{Atomisation}$	0.0
	J $K^{-1}mol^{-1}$
Standard Entropy, $S^{\ominus}$	169.7

Associated Bond Lengths (nm) & Energies (kJ mol^{-1})	Xe–O(0.176,84), Xe– F(0.194,133)

Ytterbium

[After Ytterby in Sweden]

Atomic Number	70
Relative Atomic Mass	173.04

Chemical Symbol	Yb	Group	–	Period	6 Lanthanide series

Main Source & History

Discovered by J.C.G. de Marignac in 1878, Geneva, Switzerland. The main ores are euxenite and gadolinite from which Ytterbium is extracted by ion exchange.

Properties

Ytterbium is a soft, silvery–white metal. It is both malleable and ductile. Ytterbium is a relatively unreactive element which reacts slowly with air and water from which it liberates hydrogen. It also reacts with acids.

Uses

Silicates of ytterbium have a use as synthetic gemstones.

Biological Role

Ytterbium as no known biological role.

Melting Point/K	1097.0	Boiling Point/K	1700.0	Density/g cm^{-3}	6.98

Electronic Configuration	[Xe] $4f^{14}$, $6s^2$

Oxidation State(s)	0, +2, <u>+3</u>	Isotopes	^{174}Yb(31.8%), ^{172}Yb(21.9%)

S.E.P.	$E^\ominus$/Volts
Yb^{3+}|Yb	−2.22 Acid solution

Electronegativity (Pauling)	1.2

Radii	nm
Ionic	0.086(+3)
Atomic	0.194
Van der Waals	–
Covalent	0.174

Enthalpy Data	kJ mol^{-1}
1st Ionisation Energy	603
2nd Ionisation Energy	1174
1st Electron Affinity	$+50.0^2$
2nd Electron Affinity	–
Enthalpy of Fusion, $\Delta H^\ominus_{Fusion}$	9.2
Enthalpy of Vaporisation, $\Delta H^\ominus_{Vaporisation}$	159.0
Enthalpy of Atomisation, $\Delta H^\ominus_{Atomisation}$	–
	J $K^{-1}mol^{-1}$
Standard Entropy, $S^\ominus$	59.9

Associated Bond Lengths (nm) & Energies (kJ mol^{-1})	n.a.

Yttrium

[After Ytterby in Sweden]

Atomic Number	39
Relative Atomic Mass	88.9059

Chemical Symbol	Y	Group	–	Period	5 2nd series T.M.

Main Source & History

Discovered by J. Gadolin in 1794, Finland. The main ore is gadolinite. Yttrium is extracted, with difficulty, because of its rareness. The metal is obtained by reduction of the fluoride with calcium.

Properties

Yttrium is a silvery–white metal which is malleable and ductile. It is a fairly reactive element which reacts readily with oxygen and water liberating hydrogen.

Uses	Compounds of yttrium are used as microwave filters (mixed silicates) and as red phosphors in colour televisions (YVO_4:Eu).

Biological Role	Yttrium has no known biological role.

Melting Point/K	1770.0[2]	Boiling Point/K	3611.0[2]	Density/g cm^{-3}	4.47

Electronic Configuration	[Kr] $4d^1$, $5s^2$

Oxidation State(s)	0,±3	Isotopes	^{89}Y(100%)

S.E.P.	$E^{\ominus}$/Volts
Y^{3+}|Y	– 2.37 Acid solution

Electronegativity (Pauling)	1.2

Radii	nm
Ionic	0.090 (+3)
Atomic	0.181
Van der Waals	–
Covalent	0.162

Enthalpy Data	kJ mol^{-1}
1st Ionisation Energy	616
2nd Ionisation Energy	1181
1st Electron Affinity	+39.0
2nd Electron Affinity	–
Enthalpy of Fusion, $\Delta H^{\ominus}_{Fusion}$	17.2
Enthalpy of Vaporisation, $\Delta H^{\ominus}_{Vaporisation}$	393.3
Enthalpy of Atomisation, $\Delta H^{\ominus}_{Atomisation}$	421.3
	J $K^{-1}mol^{-1}$
Standard Entropy, $S^{\ominus}$	44.4

Associated Bond Lengths (nm) & Energies (kJ mol^{-1})	n.a.

Zinc

[German, 'zink' = zinc]

Atomic Number	30
Relative Atomic Mass	65.39

Chemical Symbol	Zn	Group	–	Period	4 (1st series T.M.)

Main Source & History

Zinc was known of circa. 500BC. W. Homberg in 1695 extracted it from calamine. The principal ores of zinc are zincite (ZnO), zinc blende (ZnS) and calamine ($ZnCO_3$). The metal is obtained by roasting the ore in air followed by the reduction of the oxide with carbon.

Properties

Zinc is a bluish–white metal which is brittle at ordinary temperatures. It is a good conductor of heat and electricity. Zinc is a relatively reactive metal. It slowly reacts with oxygen and water but more vigorously with acids from which hydrogen is liberated.

Uses	Zinc is used in the production of alloys (eg. brass). It is used in galvanising steel, and for electrodes in batteries. Compounds of zinc are used in the manufacture of paints (ZnS), plastics, batteries ($ZnCl_2$) and rubbers (ZnO). They are also used as catalysts (ZnR_2–in the manufacture of polymers), mordants ($ZnSO_4$), in medicines ($ZnCO_3$), in zinc plating ($ZnSO_4$) and wood preservatives ($ZnCl_2$). The phosphor Zn_2SiO_4:Mn is important in the manufacture of devices which provide night vision.
Biological Role	Zinc is an essential trace element. It is a co–factor of many enzymes such as lactic dehydrogenase. High concentrations of zinc and its compounds are known to be toxic and thought to be carcinogenic. The recommended limit is 50 ppm (all foods). In plants zinc ions activate carboxylases. Leaves may be malformed if zinc deficiency is present.

Melting Point/K	693.0	Boiling Point/K	1180.0	Density/g cm^{-3}	7.14

Electronic Configuration	[Ar] $3d^{10}, 4s^2$

Oxidation State(s)	0,±2	Isotopes	^{64}Zn(48.6%), ^{66}Zn(27.9%)

S.E.P.	$E^{\ominus}$/Volts
Zn^{2+}|Zn	– 0.76 (Acid solution)

Electronegativity (Pauling)	1.6

Radii	nm
Ionic	0.074 (+2)
Atomic	0.133
Van der Waals	–
Covalent	0.125

Enthalpy Data	kJ mol^{-1}
1st Ionisation Energy	906
2nd Ionisation Energy	1733
1st Electron Affinity	+9.0
2nd Electron Affinity	–
Enthalpy of Fusion, $\Delta H^{\ominus}_{Fusion}$	7.4
Enthalpy of Vaporisation, $\Delta H^{\ominus}_{Vaporisation}$	115.3
Enthalpy of Atomisation, $\Delta H^{\ominus}_{Atomisation}$	130.7
	J $K^{-1}mol^{-1}$
Standard Entropy, $S^{\ominus}$	41.6

Associated Bond Lengths (nm) & Energies (kJ mol^{-1})	n.a.

Zirconium

[Persian, 'zargun' = gold coloured]

Atomic Number	40
Relative Atomic Mass	91.224

Chemical Symbol	Zr	Group	–	Period	5 2nd series T.M.

Main Source & History

Discovered by M.H. Klaproth in 1788, Berlin, Germany. The principal ores are baddeleyite (ZrO_2) and zircon ($ZrSiO_4$). The metal is obtained by reduction of the oxide or chloride by calcium or magnesium (Kroll Process).

Properties

Zirconium is a hard, silvery–white metal, lustrous in appearance. It is a relatively unreactive element extremely resistant to corrosion and unaffected by acids.

Uses

Zirconium is used extensively in the production of alloys for reactor construction and superconducting magnets (Nb–Zr). Compounds of zirconium are used for flame proofing and as abrasives, also as refractories (ZrO_2) and pigments (ZrO_2).

Biological Role

Zirconium has no known biological role although it is thought that zirconium and its compounds may be carcinogenic.

Melting Point/K	2125.0	Boiling Point/K	4650.0[2]	Density/g cm^{-3}	6.49

Electronic Configuration	[Kr] $4d^2$, $5s^2$

Oxidation State(s)	0, +2, +3, $\underline{+4}$	Isotopes	^{90}Zr(51.45%), ^{94}Zr(17.4%)

S.E.P.	$E^{\ominus}$/Volts
Zr^{4+}|Zr	– 1.55

Electronegativity (Pauling)	1.4

Radii	nm
Ionic	0.080 (+4)
Atomic	0.160
Van der Waals	–
Covalent	0.145

Enthalpy Data	kJ mol^{-1}
1st Ionisation Energy	660
2nd Ionisation Energy	1267
1st Electron Affinity	+43.0
2nd Electron Affinity	–
Enthalpy of Fusion, $\Delta H^{\ominus}_{Fusion}$	16.7
Enthalpy of Vaporisation, $\Delta H^{\ominus}_{Vaporisation}$	566.7
Enthalpy of Atomisation, $\Delta H^{\ominus}_{Atomisation}$	608.8
	J $K^{-1}mol^{-1}$
Standard Entropy, $S^{\ominus}$	39.0

Associated Bond Lengths (nm) & Energies (kJ mol^{-1})	n.a.

The following pages contain graphs showing some of the numerical data for the first thirty six elements of the Periodic Table. These graphs are included to show the general periodic trends related to certain properties of these elements.

Graph to show 1st ionisation energy for the first 36 elements

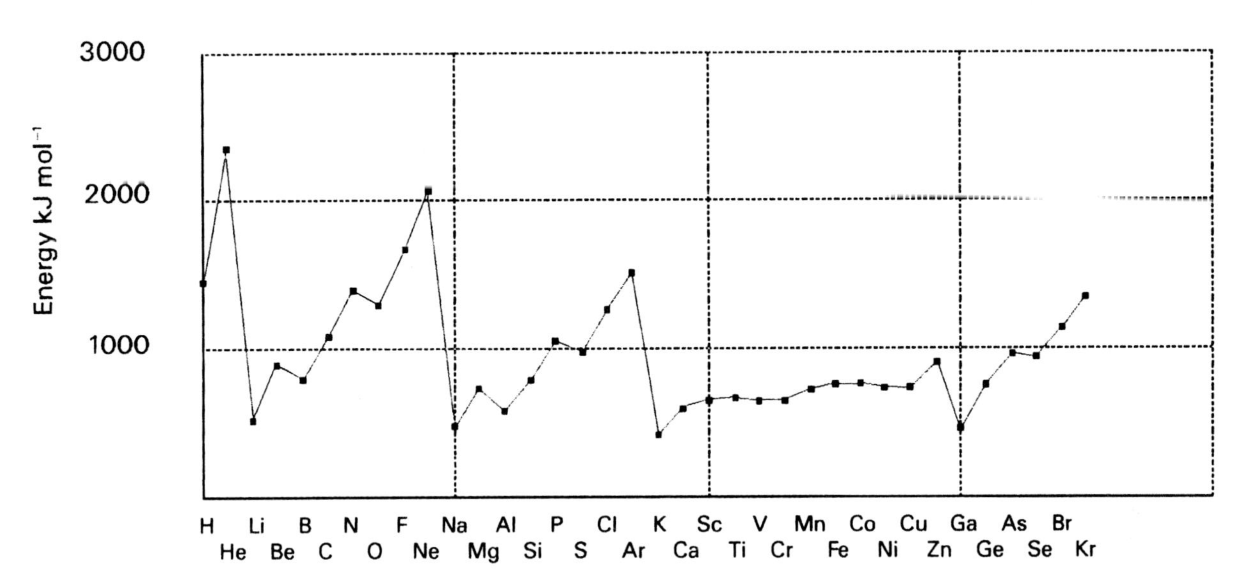

Graph to show electronegativity for the first 36 elements

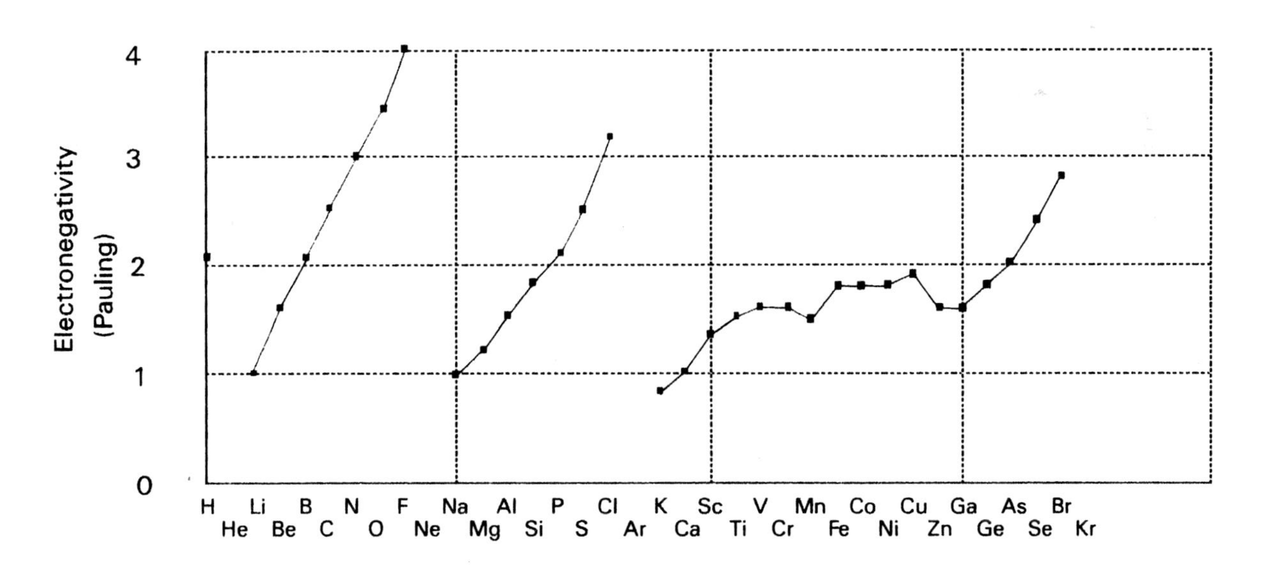

Graph to show standard molar enthalpy of fusion for the first 36 elements

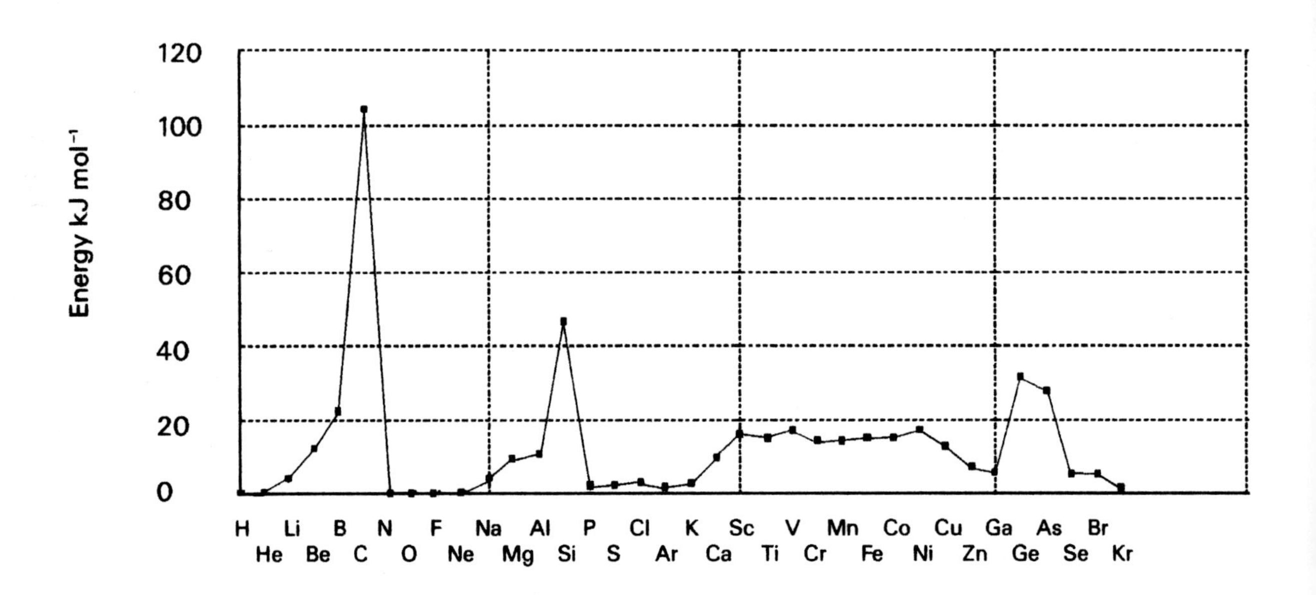

Graph to show melting point for the first 36 elements

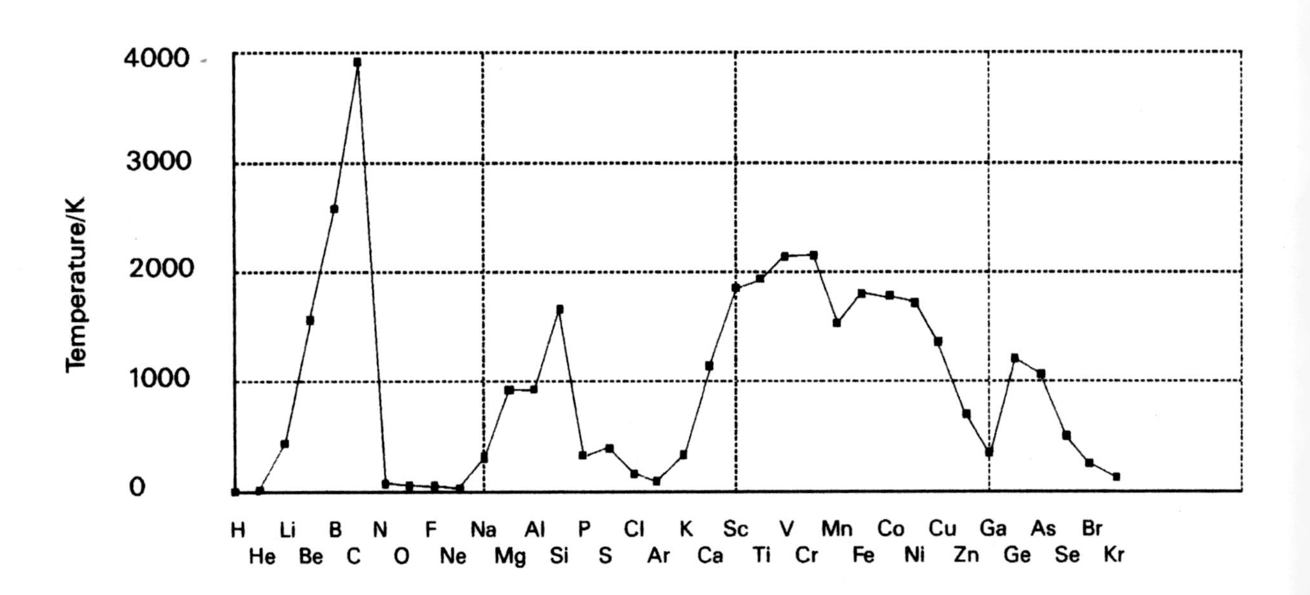

Graph to show standard molar enthalpy of vaporisation for the first 36 elements

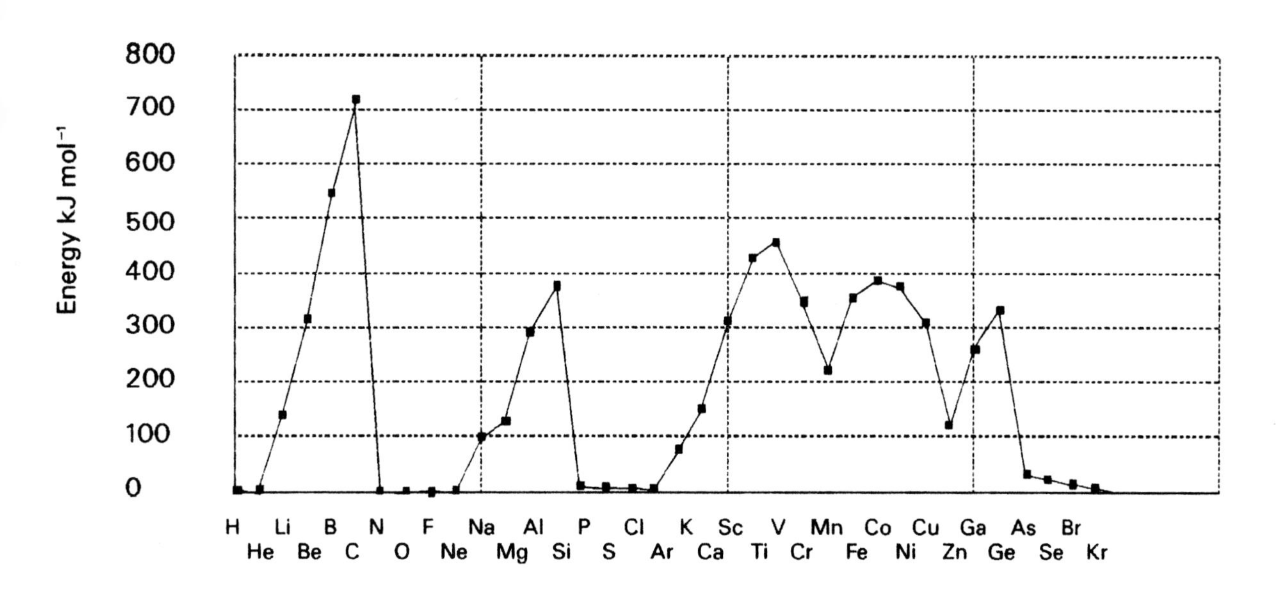

Graph to show boiling point for the first 36 elements

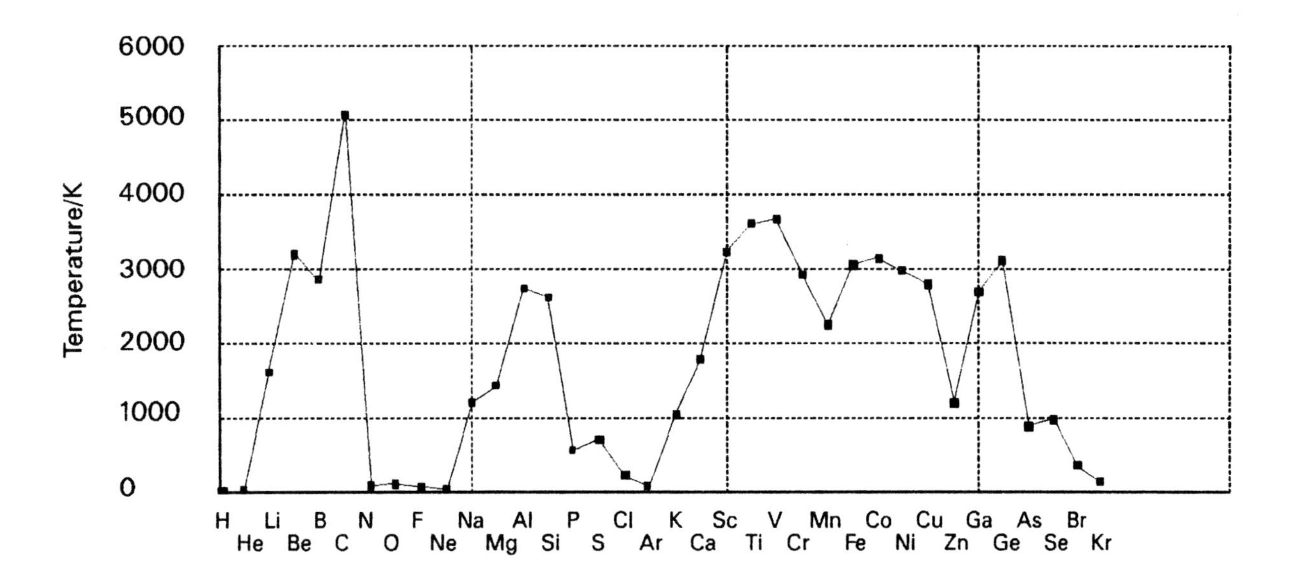

Graph to show the atomic radius for the first 36 elements

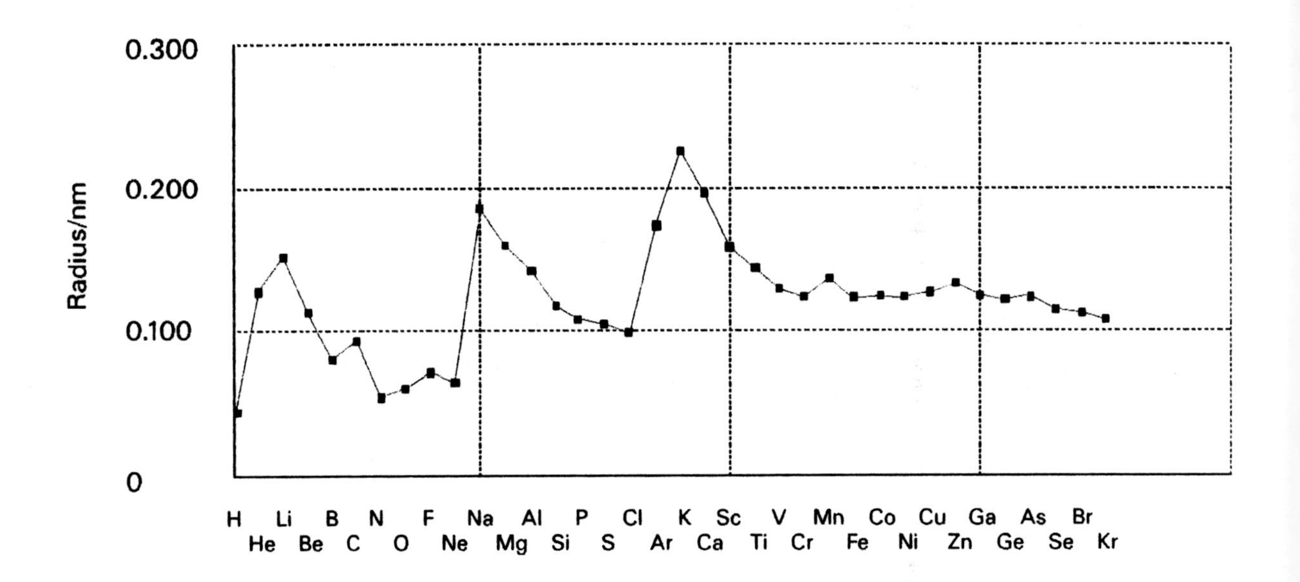

Graph to show the density of the first 36 elements

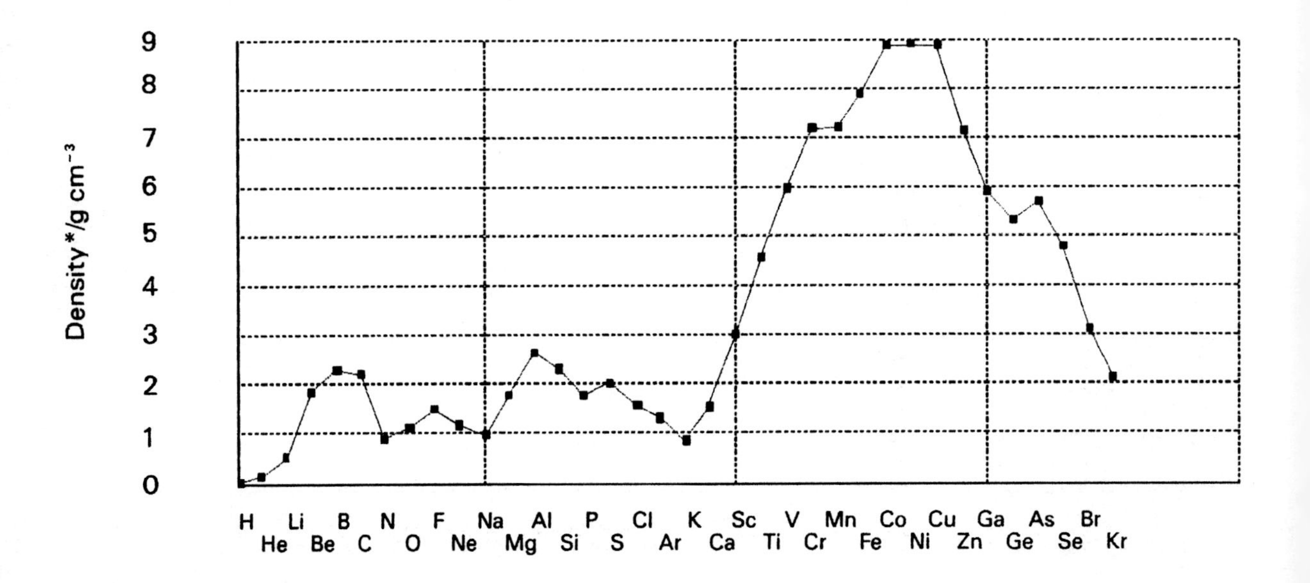

* Density of gases measured at their boiling point.

COMPOUNDS

USEFUL DATA

Glossary of terms and units

This section contains tabulated data related to some common inorganic and organic compounds. The information given for each compound is sub–divided into the following categories.

a) The physical state of the substance at 298K and 1 atmosphere pressure: s = solid, l = liquid and g = gas.

b) The melting and boiling points of the compounds measured in Kelvin (K).

c) The density measured in g cm^{-3} at 298K and 1 atmosphere pressure except where stated.

d) The standard molar enthalpy of combustion ($\Delta H^{\ominus}_c$) of a substance is defined as the enthalpy change which takes place when one mole of that substance, in its standard state, is completely burnt in oxygen at 298K and 1 atmosphere pressure. $\Delta H^{\ominus}_c$ is measured in kilojoules per mole (kJ mol^{-1}).

e) The standard molar enthalpy of formation ($\Delta H^{\ominus}_f$) of a substance is defined as the enthalpy change which takes place when one mole of that substance is formed from its constituent elements, in their standard states, at 298K and 1 atmosphere pressure. $\Delta H^{\ominus}_f$ is measured in kilojoules per mole (kJ mol^{-1}).

f) The standard molar Gibbs free energy of formation ($\Delta G^{\ominus}_f$) of a substance is defined as the free energy change that accompanies the formation of one mole of that substance from its elements in their standard states, at 298K and 1 atmosphere pressure. $\Delta G^{\ominus}_f$ is measured in kilojoules per mole (kJ mol^{-1}). This value gives an indication of the stability of the substance.

g) The standard molar entropy ($S^{\ominus}$) is defined as the entropy of 1 mole of a substance in its standard state at 298K and 1 atmosphere pressure. This value gives a measure of the degree of disorder. In general $S^{\ominus}_{(g)} > S^{\ominus}_{(l)} > S^{\ominus}_{(s)}$. $S^{\ominus}$ is measured in Joules per Kelvin per mole ($J\ K^{-1}\ mol^{-1}$).

h) The solubility of the substance in water (Sol_{SAT}) is expressed as the mass of solute (grams) required to saturate 100 grams of water at 298K and 1 atmosphere pressure. Sol_{SAT} is measured in g/100g H_2O.

i) The standard lattice enthalpy (L.E.) of an ionic substance is defined as the enthalpy change which takes placewhen one mole of that substance is formed from its gaseous ions. This term relates to the exothermicprocess,

$$mM^{n+}_{(g)} + nX^{m-}_{(g)} \longrightarrow M_mX_{n(s)}$$

at 298K and 1 atmosphere pressure. L.E. is measured in kilojoules per mole (kJ mol^{-1}).

j) The solubility product (K_{SP}) of the sparingly soluble salt, M_mX_n, is the product of the concentrations of the ions in a saturated solution of that salt, raised to the appropriate powers eg. $K_{SP} = [M^{n+}]^m[X^{m-}]^n$ at 298K and 1 atmosphere pressure. The units of K_{SP} vary according to :units = $mol^{(m+n)}\ dm^{-3(m+n)}$.

k) pK_{in}. Indicators are usually weak acids or weak bases which change colour with pH. For a weak acid indicator, HIn, in aqueous solution:

$$HIn_{(aq)} \rightleftharpoons H^+_{(aq)} + In^-_{(aq)}$$

$$K_{in} = \frac{[H^+][In^-]}{[HIn]}$$

(K_{in} = indicator dissociation constant)

At the end point, $[HIn] = [In^-]$

Therefore at this point,

$$K_{in} = [H^+] \qquad ([H^+] \text{ is measured in mol dm}^{-3})$$

hence

$$-\log_{10} K_{in} = -\log_{10}[H^+]$$

or

$$\underline{pK_{in} = pH}$$

l) The acid dissociation constant, K_a, for the weak acid $HA_{(aq)}$, for the following equilibrium:

$$HA_{(aq)} \rightleftharpoons H^+_{(aq)} + A^-_{(aq)}$$

is given by

$$K_a = \frac{[H^+][A^-]}{[HA]}$$

K_a has units of mol dm^{-3}.

m) Infrared data. Infrared radiation occupies the wavelength range from about 25000nm (or 400 cm^{-1}) to 2500 nm (or 4000 cm^{-1}). All organic molecules absorb strongly in the infrared causing molecules to rotate and bonds to vibrate. The wavelengths absorbed by bonds stretching and bending in a particular functional group is characteristic of that group. Measurements made using infrared spectroscopy provide information about the structural formula of organic molecules.

a) Inorganic compounds: selected physical and thermochemical data.

M.Pt. = Melting point at 1 atm except where stated

B.Pt. = Boiling point at 1 atm except where stated

$\Delta H^{\ominus}_{f}$ Standard molar enthalpy change of formation at 298K and 1 atmosphere pressure.

$\Delta G^{\ominus}_{f}$ = Standard molar Gibbs free energy of formation at 298K and 1 atmosphere pressure.

$S^{\ominus}$ = Standard molar entropy at 298K and 1 atmosphere pressure.

Sol_{SAT} = Solubility in water, measured in g/100g of water at 298K and 1 atm pressure except where stated.

L.E. = Standard lattice enthalpy measured at 298K and 1 atmosphere pressure. In certain instances the Born–Haber cycle value (BH) is given.

K_{SP} = Solubility product for M_mX_n at 298K. Units = $mol^{(m+n)}\ dm^{-3(m+n)}$.

	Formula	M.Pt. K	B.Pt. K	$\Delta H^{\ominus}_{f}$ kJ mol^{-1}	$\Delta G^{\ominus}_{f}$ kJ mol^{-1}	$S^{\ominus}$ J K^{-1}mol^{-1}	Sol_{SAT} g/100g water	L.E. kJ mol^{-1}	K_{SP}
Aluminium									
Aluminium ion	$Al^{3+}_{(aq)}$	–	–	–531	–485	–32	–	–	–
Aluminium chloride	$AlCl_{3(s)}$	451^{sub}	–	–695	–629	+170	69.3	–	–
Aluminium hydroxide	$Al(OH)_{3(s)}$	523^{deh}	–	–1273	–1150	+85	1.0×10^{-6}	–	1×10^{-32}
Aluminium nitrate	$Al(NO_3)_3$ †	–	–	–2851	–2204	+468	–	–	–
Aluminium oxide	$Al_2O_{3(s)}$	2290	3250	–1676	–1582	+51	1.0×10^{-8}	–	–
Aluminium sulphate	$Al_2(SO_4)_{3(s)}$	1043^{dec}	–	–3441	–3100	+239	31.3	–	–
† = $.6H_2O$									
Barium									
Barium ion	$Ba^{2+}_{(aq)}$	–	–	–538	–561	+10	–	–	–
Barium carbonate	$BaCO_{3(s)}$	1123^{dec}	–	–1219	–1138	+112	1.8×10^{-3}	–	5.1×10^{-9}
Barium chloride	$BaCl_{2(s)}$	1240	1820	–859	–810	+124	36	–2056	–
Barium chromate(VI)	$BaCrO_{4(s)}$	–	–	–1428	–1339	+152	2.9×10^{-4}	–	1.2×10^{-10}
Barium nitrate	$Ba(NO_3)_{2(s)}$	865	dec	–992	–797	+214	10.2	–	–
Barium sulphate	$BaSO_{4(s)}$	1853	–	–1465	–1353	+132	2.2×10^{-4}	–	1.3×10^{-10}
Boron									
Boron Trifluoride	$BF_{3(g)}$	129	173	–1137	–1120	+254	0.32^{273K}	–	–
Calcium									
Calcium ion	$Ca^{2+}_{(aq)}$	–	–	–543	–554	–53	–	–	–
Calcium carbonate	$CaCO_{3(s)}$	$1612^{1025atm}$	1172^{dec}	–1207	–1129	+93	1.3×10^{-3}	–	5×10^{-9}
Calcium chloride	$CaCl_{2(s)}$	1050	1900	–795	–750	+114	59.5	–2237	–
Calcium hydroxide	$Ca(OH)_{2(s)}$	853^{deh}	dec	–987	–897	+76	0.11	–	5.5×10^{-6}
Calcium nitrate	$Ca(NO_3)_{2(s)}$	834	–	–938	–743	+129	102	–	–
Calcium oxide	$CaO_{(s)}$	2850	3120	–635	–604	+40	0.13	–3401	–
Calcium sulphate	$CaSO_{4(s)}$	1723	–	–1433	–1320	+107	0.63	–	2×10^{-5}
Carbon									
Carbon dioxide	$CO_{2(g)}$	162	195	–394	–394	+214	0.14^{2}	–	–
Carbon monoxide	$CO_{(g)}$	74	84	–111	–137	+198	6×10^{-42}	–	–
Hydrogen cyanide	$HCN_{(g)}$	260	299	+130	+120	+202	1.22	–	–
Chlorine									
Chloride ion	$Cl^{-}_{(aq)}$	–	–	–167	–131	+57	–	–	–
Hydrogen chloride	$HCl_{(g)}$	158	188	–92	–95	+187	217.9	–	–
Chromium									
Chromium ion	$Cr^{3+}_{(aq)}$	–	–	–232	–205	–	–	–	–
Chromium(III) hydroxide	$Cr(OH)_{3(s)}$	–	–	–1033	–	–	–	–	1×10^{-30}

	Formula	M.Pt. K	B.Pt. K	$\Delta H^{\ominus}_f$ kJ mol⁻¹	$\Delta G^{\ominus}_f$ kJ mol⁻¹	$S^{\ominus}$ J K⁻¹mol⁻¹	Sol_{SAT} g/100g water	L.E. kJ mol⁻¹	K_{SP}
Chromium(III) oxide	$Cr_2O_{3(s)}$	2538	4273	−1128	−1047	+81	1.8×10^{-7}	–	–
Cobalt									
Cobalt(II) ion	$Co^{2+}_{(aq)}$	–	–	−58	−54	−113	–	–	–
Cobalt(II) chloride	$CoCl_{2(s)}$	997 in HCl	1322	−326	−270	+109	44	–	–
Cobalt(II) hydroxide	$Co(OH)_{2(s)}$	dec	–	−549	−454	+79	1.3×10^{-4}	–	6.3×10^{-16}
Cobalt(II) nitrate	$Co(NO_3)_{2(s)}$	–	–	−421	−237	+192	–	–	–
Cobalt(II) oxide	$CoO_{(s)}$	2208	–	−239	−214	+53	ins	–	–
Cobalt(II) sulphate	$CoSO_{4(s)}$	1008 dec	–	−888	−782	+113	36.3	–	–
Copper									
Copper(I) ion	$Cu^{+}_{(aq)}$	–	–	+72	+50	+41	–	–	–
Copper(II) ion	$Cu^{2+}_{(aq)}$	–	–	+65	+66	−100	–	–	–
Copper(I) chloride	$CuCl_{(s)}$	703	1763	−137	−120	+86	6×10^{-3}	−976	–
Copper(II) chloride	$CuCl_{2(s)}$	893	1266 dec	−220	−176	+108	73	−993	–
Copper(II) hydroxide	$Cu(OH)_{2(s)}$	dec	–	−448	−360	+75	–	–	2×10^{-19}
Copper(I) iodide	$CuI_{(s)}$	878	1503	−68	−70	+97	8×10^{-4}	–	–
Copper(II) nitrate	$Cu(NO_3)_{2(s)}$	–	–	−307	−118	+193	122	–	–
Copper(I) oxide	$Cu_2O_{(s)}$	1508	2073 dec	−167	−146	+101	ins	–	–
Copper(II) oxide	$CuO_{(s)}$	1599	–	−155	−127	+44	2.4×10^{-4}	−3189	–
Copper(II) sulphate (anh)	$CuSO_{4(s)}$	473	923 dec	−770	−661	+113	20.5	–	–
Copper(II) sulphate (hyd)	$CuSO_4$ †	383 deh	423 deh	−2278	−1880	+300	34.7	–	–
† = $.5H_2O$									
Fluorine									
Fluoride ion	$F^{-}_{(aq)}$	–	–	−333	−279	−14	–	–	–
Hydrogen fluoride	$HF_{(g)}$	190	293	−269	−273	+174	0.87	–	–
Hydrogen									
Hydrogen ion	$H^{+}_{(aq)}$	–	–	0	0	0	–	–	–
Water (gas)	$H_2O_{(g)}$	273	373	−242	−229	+189	–	–	–
Water (liquid)	$H_2O_{(l)}$	273	373	−286	−237	+70	–	–	–
Hydrogen peroxide	$H_2O_{2(l)}$	273	323	−188	−118	+110	infinite	–	–
Iodine									
Iodide ion	$I^{-}_{(aq)}$	–	–	−55	−52	+111		–	–
Hydrogen iodide	$HI_{(g)}$	222	238	+26	+2	+207	0.0002atm 7.1	–	–
Iron									
Iron(II) ion	$Fe^{2+}_{(aq)}$	–	–	−89	−79	−138	–	–	–
Iron(III) ion	$Fe^{3+}_{(aq)}$	–	–	−49	−5	−316	–	–	–
Iron(II) chloride	$FeCl_{2(s)}$	954	–	−342	−302	+120	64.4	–	–
Iron(III) chloride	$FeCl_{3(s)}$	579 sub	588 dec	−405	−334	+142	dec	–	–
Iron(II) hydroxide	$Fe(OH)_{2(s)}$	dec	–	−568	−487	+88	6×10^{-4}	–	6×10^{-15}
Iron(III) hydroxide	$Fe(OH)_{3(s)}$	–	–	−824	−697	+107	3.6×10^{-5}	–	8×10^{-40}
Iron(II) oxide	$FeO_{(s)}$	1642	–	−267	−244	+59	–	–	–
Iron(III) oxide	$Fe_2O_{3(s)}$	1838	–	−822	−741	+90	–	–	–
Iron(II) sulphate	$FeSO_{4(s)}$	–	–	−923	−821	+108	15.6	–	–
Lead									
Lead(II) ion	$Pb^{2+}_{(aq)}$	–	–	−2	−24	+11	–	–	–
Lead(II) carbonate	$PbCO_{3(s)}$	588 dec	–	−700	−626	+131	1.1×10^{-4}	–	6.3×10^{-14}
Lead(II) chloride	$PbCl_{2(s)}$	774	1220	−359	−314	+136	1.1	−2269	2×10^{-5}
Lead(IV) chloride	$PbCl_{4(s)}$	258	378 exp	−329	−259	–	dec	–	–
Lead(II) hydroxide	$Pb(OH)_{2(s)}$	–	–	−515	−421	+88	1.6×10^{-2}	–	–
Lead(II) nitrate	$Pb(NO_3)_{2(s)}$	743	–	−449	−251	+213	148	–	–
Lead(II) oxide	$PbO_{(s)}$	1161 dec	1745	−218	−188	+69	2.4×10^{-3}	−3502 BH	–
Lead(IV) oxide	$PbO_{2(s)}$	560	–	−277	−217	+69	–	–	–
Lead(II) sulphate	$PbSO_{4(s)}$	1443 dec	–	−918	−811	+149	4.5×10^{-3}	–	1.6×10^{-8}

	Formula	M.Pt. K	B.Pt. K	$\Delta H^{\ominus}_f$ kJ mol^{-1}	$\Delta G^{\ominus}_f$ kJ mol^{-1}	$S^{\ominus}$ J K^{-1}mol^{-1}	Sol$_{SAT}$ g/100g water	L.E. kJ mol^{-1}	K_{SP}
Lithium									
Lithium ion	$Li^+_{(aq)}$	–	–	−279	−293	+13	–	–	–
Lithium carbonate	$Li_2CO_{3(s)}$	996	1583^{dec}	−1216	−1133	+90	1.29	–	–
Lithium chloride	$LiCl_{(s)}$	887	1600	−409	−384	+59	85	−848	–
Lithium nitrate	$LiNO_{3(s)}$	537	873^{dec}	−482	−381	+90	70	–	–
Lithium oxide	$Li_2O_{(s)}$	973^2	–	−596	−560	+38	dec	−2814	–
Lithium sulphate	$Li_2SO_{4(s)}$	–	1118	−1434	−1322	+115	25.9	–	–
Magnesium									
Magnesium ion	$Mg^{2+}_{(aq)}$	–	–	−467	−455	−138	–	–	–
Magnesium carbonate	$MgCO_{3(s)}$	620	1173^{dec}	−1113	−1012	+66	0.01	–	1×10^{-5}
Magnesium chloride	$MgCl_{2(s)}$	981	1685	−642	−592	+90	53	−2526	–
Magnesium hydroxide	$Mg(OH)_{2(s)}$	620^{deh}	–	−925	−834	+63	1.2×10^{-3}	–	2×10^{-11}
Magnesium nitrate	$Mg(NO_3)_{2(s)}$	–	–	−790	−590	+104	70	–	–
Magnesium nitride	$Mg_3N_{2(s)}$	1073^{dec}	–	−461	−406	+90	dec	–	–
Magnesium oxide	$MgO_{(s)}$	3100	3900	−602	−570	+27	8.1×10^{-4}	−3889	–
Magnesium sulphate	$MgSO_{4(s)}$	1397^{dec}	–	−1278	−1171	+92	22	–	–
Manganese									
Manganate(VII) ion	$MnO_4{}^-_{(aq)}$	–	–	−541	−447	+191	–	–	–
Manganese(II) ion	$Mn^{2+}_{(aq)}$	–	–	−221	−228	−74	–	–	–
Manganese(II) hydroxide	$Mn(OH)_{2(s)}$	dec	–	−694	−610	+99	2×10^{-4}	–	2×10^{-13}
Manganese(IV) oxide	$MnO_{2(s)}$	808^{dec}	–	−521	−466	+53	ins	–	–
Nickel									
Nickel(II) ion	$Ni^{2+}_{(aq)}$	–	–	−54	−46	−129	–	–	–
Nickel(II) oxide	$NiO_{(s)}$	2260	–	−244	−216	+39	ins	–	–
Nickel(II) sulphate	$NiSO_{4(s)}$	1121^{dec}	–	−891	−774	+79	29.3	–	–
Nitrogen									
Ammonium ion	$NH_4{}^+_{(aq)}$	–	–	−133	−79	+113	–	–	–
Ammonia	$NH_{3(g)}$	195	240	−46	−17	+193	52.9	–	–
Ammonium chloride	$NH_4Cl_{(s)}$	613^{sub}	–	−315	−203	+95	39.3	−705	–
Ammonium hydroxide	$NH_4OH_{(l)}$	–	–	−361	−254	+166	–	–	–
Ammonium nitrate	$NH_4NO_{3(s)}$	443	483	−366	−184	+151	214.4	–	–
Ammonium sulphate	$(NH_4)_2SO_{4(s)}$	508^{dec}	–	−1179	−902	+220	76.4	–	–
Nitrogen(I) oxide	$N_2O_{(g)}$	182	185	+82	+104	+220	0.12	–	–
Nitrogen(II) oxide	$NO_{(g)}$	110	121	+90	+87	+211	5.6×10^{-3}	–	–
Nitrogen(IV) oxide	$NO_{2(g)}$	262^{eqm}	294^{eqm}	+34	+52	+240	sol	–	–
Dinitrogen tetroxide	$N_2O_{4(g)}$	262^{eqm}	294^{eqm}	+10	+98	+304	dec	–	–
Nitrogen(V) oxide	$N_2O_{5(s)}$	303	320^{dec}	−42	+114	+178	dec	–	–
Oxygen									
Ozone	$O_{3(g)}$	81	161	+143	+163	+239	10.5	–	–
Phosphorus									
Phosphine	$PH_{3(g)}$	140	185	+5	+13	+210	0.03	–	–
Phosphoric(V) acid	$H_3PO_{4(s)}$	316	486^{deh}	−1281	−1119	+111	669.3	–	–
Phosphorus(III) chloride	$PCl_{3(l)}$	161	349	−320	−272	+217	dec	–	–
Phosphorus(V) chloride	$PCl_{5(s)}$	435^{sub}	–	−463	–	+167	dec	–	–
Potassium									
Potassium ion	$K^+_{(aq)}$	–	–	−252	−283	+103	–	–	–
Potassium bromide	$KBr_{(s)}$	1007	1708	−392	−381	+96	68	−670	–
Potassium carbonate	$K_2CO_{3(s)}$	1164	dec	−1146	−1064	+156	112.1	–	–
Potassium chloride	$KCl_{(s)}$	1049	1770^{sub}	−436	−409	+83	35.9	−711	–
Potassium hydrogencarbonate	$KHCO_{3(s)}$	400^{dec}	–	−959	−864	+116	36.2	–	–
Potassium hydroxide	$KOH_{(s)}$	633	1503	−426	−379	+79	96	–	–

	Formula	M.Pt. K	B.Pt. K	$\Delta H^\ominus_f$ kJ mol^{-1}	$\Delta G^\ominus_f$ kJ mol^{-1}	$S^\ominus$ J K^{-1}mol^{-1}	Sol$_{SAT}$ g/100g water	L.E. kJ mol^{-1}	K_{SP}
Potassium iodide	$KI_{(s)}$	954	1603	−328	−325	+106	148.1	−629	–
Potassium nitrate	$KNO_{3(s)}$	607	673^{dec}	−493	−393	+133	37.9	–	–
Potassium sulphate	$K_2SO_{4(s)}$	1342	1962	−1434	−1316	+176	12	–	–
<u>Silicon</u>									
Silane	$SiH_{4(g)}$	88	161	+34	+57	+205	ins	–	–
Silicon(IV) oxide	$SiO_{2(s)}$	1880	2500	−911	−857	+42	0.01	–	–
<u>Silver</u>									
Silver ion	$Ag^+_{(aq)}$	–	–	+106	+77	+73	–	–	–
Silver bromide	$AgBr_{(s)}$	705	1573^{dec}	−100	−94	+107	1.4×10^{-5}	−890	5×10^{-13}
Silver carbonate	$Ag_2CO_{3(s)}$	491^{dec}	–	−506	−437	+167	0.03	–	6.3×10^{-12}
Silver chloride	$AgCl_{(s)}$	728	1823	−127	−110	+96	1.9×10^{-4}	−905	2×10^{-10}
Silver chromate(VI)	$Ag_2CrO_{4(s)}$	–	–	−712	−622	+217	0.03	–	3×10^{-12}
Silver fluoride	$AgF_{(s)}$	708	1432	−203	−185	+80	180.2	−955	–
Silver iodide	$AgI_{(s)}$	831	1779	−62	−66	+114	2.6×10^{-6}	−876	8×10^{-17}
Silver nitrate	$AgNO_{3(s)}$	485	1717^{dec}	−123	−32	+141	241.3	–	–
Silver oxide	$Ag_2O_{(s)}$	503^{dec}	–	−31	−11	+122	4.6×10^{-3}	−2910	–
Silver sulphate	$Ag_2SO_{4(s)}$	925	1358	−716	−619	+200	0.6	–	1.7×10^{-5}
<u>Sodium</u>									
Sodium ion	$Na^+_{(aq)}$	–	–	−240	−262	+321	–	–	–
Sodium chloride	$NaCl_{(s)}$	1074	1686	−411	−384	+72	35.9	−780	–
Sodium bromide	$NaBr_{(s)}$	1028	1660	−360	−349	+87	94.6	−742	–
Sodium iodide	$NaI_{(s)}$	933	1577	−288	−286	+99	184.4	−705	–
Sodium oxide	$Na_2O_{(s)}$	1548^{sub}	–	−416	−377	+73	dec	−2478	–
Sodium peroxide	$Na_2O_{2(s)}$	733^{dec}	–	−511	−448	+95	dec	–	–
Sodium hydrogencarbonate	$NaHCO_{3(s)}$	540^{dec}	–	−948	−852	+102	10.2	–	–
Sodium hydroxide	$NaOH_{(s)}$	592	1660	−427	−380	+65	42	–	–
Sodium carbonate	$Na_2CO_{3(s)}$	1124	dec	−1131	−1048	+136	7	–	–
Sodium nitrate(III) (nitrite)	$NaNO_{2(s)}$	544	593^{dec}	−359	−285	+104	84.9	–	–
Sodium nitrate(V)	$NaNO_{3(s)}$	580	653^{dec}	−468	−367	+117	91.9	–	–
Sodium sulphate	$Na_2SO_{4(s)}$	1157	–	−1385	−1267	+150	4.3	–	–
Sodium thiosulphate	$Na_2S_2O_{3(s)}$	–	–	−1123	−1028	+155	50	–	–
<u>Sulphur</u>									
Sulphate ion	$SO_4^{2-}{}_{(aq)}$	–	–	−909	−745	+20	–	–	–
Sulphide ion	$S^{2-}{}_{(aq)}$	–	–	+33	+86	−15	–	–	–
Sulphur(VI) fluoride	$SF_{6(g)}$	209^{sub}	–	−1209	−1105	+292	0.54	–	–
Sulphur dioxide	$SO_{2(g)}$	200	263	−297	−300	+248	10.6	–	–
Sulphur(VI) oxide	$SO_{3(g)}$	306	318	−395	−370	+256	infinite	–	–
Sulphuric acid	$H_2SO_{4(l)}$	284	610	−814	−690	+157	infinite	–	–
<u>Tin</u>									
Tin(II) ion	$Sn^{2+}{}_{(aq)}$	–	–	−9	−27	−17	–	–	–
Tin(IV) ion	$Sn^{4+}{}_{(aq)}$	–	–	+31	+3	−117	–	–	–
Tin(II) chloride	$SnCl_{2(s)}$	519	925	−325	–	–	270	–	–
Tin(IV) chloride	$SnCl_{4(l)}$	240	387	−511	−240	+259	dec	–	–
Tin(II) oxide	$SnO_{(s)}$	1350^{dec}	–	−286	−257	+57	6.7×10^{-5}	–	–
Tin(IV) oxide	$SnO_{2(s)}$	1903	2123^{sub}	−581	−52	+52	2.1×10^{-9}	BH −11770	–
<u>Zinc</u>									
Zinc ion	$Zn^{2+}{}_{(aq)}$	–	–	−154	−147	−112	–	–	–
Zinc carbonate	$ZnCO_{3(s)}$	570^{dec}	–	−812	−732	+82	0.02	–	1.4×10^{-11}
Zinc chloride	$ZnCl_{2(s)}$	556	1005	−416	−369	+108	413	−2734	–
Zinc hydroxide	$Zn(OH)_{2(s)}$	–	–	−642	–	–	–	–	2×10^{-17}
Zinc nitrate	$Zn(NO_3)_{2(s)}$	–	–	−482	–	–	117	–	–
Zinc sulphate	$ZnSO_{4(s)}$	873^{dec}	–	−979	−872	+125	54	–	–

b) Organic compounds: selected physical and thermochemical data

State = The physical state of the substance at 298K and 1 atmosphere pressure: s = solid, l = liquid and g = gas.
M.Pt. = Melting point at 1 atmosphere except where stated.
B.Pt. = Boiling point at 1 atmosphere except where stated.
Density = The density measured in g cm^{-3} at 298K and 1 atmosphere pressure.
$\Delta H^{\ominus}_f$ = Standard molar enthalpy of formation at 298K and 1 atmosphere pressure.
$\Delta H^{\ominus}_c$ = Standard molar enthalpy of combustion at 298K and 1 atmosphere pressure.
$\Delta G^{\ominus}_f$ = Standard molar Gibbs free energy of formation at 298K and 1 atmosphere pressure.
$S^{\ominus}$ = Standard molar entropy at 298K and 1 atmosphere pressure.

		State (s,l,g)	M.Pt. K	B.Pt. K	Density g cm^{-3}	$\Delta H^{\ominus}_f$ kJ mol^{-1}	$\Delta H^{\ominus}_c$ kJ mol^{-1}	$\Delta G^{\ominus}_f$ kJ mol^{-1}	$S^{\ominus}$ J $K^{-1}mol^{-1}$
Alkanes									
Methane	CH_4	g	91.1	109.1	0.466[liq]	−74.9	−890.4	−50.8	+186.0
Ethane	CH_3CH_3	g	89.9	184.5	0.572[liq]	−84.7	−1560.0	−32.9	+230.0
Propane	$CH_3CH_2CH_3$	g	83.4	230.9	0.585[liq]	−104.0	−2220.0	−23.5	+270.0
Butane	$CH_3(CH_2)_2CH_3$	g	134.7	272.6	0.601[liq]	−146.2	−2877.0	−15.7	+310.0
Methylpropane	$CH_3(CH_2)_3CH_3$	l	143.0	309.2	0.626	−173.0	−3509.1	−8.2	+261.2
Hexane	$CH_3(CH_2)_4CH_3$	l	177.7	341.7	0.659	−198.8	−4163.0	−4.2	+295.9
Heptane	$CH_3(CH_2)_5CH_3$	l	182.5	371.5	0.684	−224.4	−4853.0	+1.3	+328.5
Octane	$CH_3(CH_2)_6CH_3$	l	216.3	398.8	0.703	−250.0	−5512.0	+6.4	+361.1
Eicosane	$CH_3(CH_2)_{18}CH_3$	s	309.9	616.9	0.789	–	–	–	–
2−methylpropane	$(CH_3)_2CHCH_3$	g	113.7	261.3	0.557[liq]	−134.6	−2868.5	−17.9	+294.6
2−methylbutane	$(CH_3)_2CHCH_2CH_3$	l	115.0	301.0	0.620	−178.9	−3503.4	−14.5	+260.4
2,2−dimethylpropane	$C(CH_3)_4$	g	256.6	282.5	0.591[liq]	−189.8	−3492.5	−15.2	+306.4
cyclohexane	$CH_2(CH_2)_4CH_2$	l	279.6	353.8	0.779	−156.2	−3924.0	+26.8	+204.0
Alkenes									
Ethene	$CH_2{=}CH_2$	g	104.0	169.0	0.610[liq]	+52.3	−1411.0	+68.1	+219.5
Propene	$CH_3CH{=}CH_2$	g	88.0	225.7	0.514[liq]	+20.4	−2058.1	+74.7	+267.0
But−1−ene	$CH_3CH_2CH{=}CH_2$	g	88.8	266.8	0.595[liq]	−0.4	−2716.8	+72.0	+307.0
Cyclohexene	$CH_2(CH_2)_3CH{=}CH_2$	l	169.6	356.4	0.810	−38.1	−3751.9	–	–
Phenylethene (styrene)	$C_6H_5CH{=}CH_2$	l	242.5	418.3	0.906	+104.0	−4395.0	+214.0	+345.1
Alkynes									
Ethyne	$CH{\equiv}CH$	g	189.4[sub]	–	0.618[liq]	+226.7	−1300.0	+209.0	+201.0
Aromatic hydrocarbons (arenes)									
Benzene	C_6H_6	l	278.6	353.1	0.878	+48.7	−3273.0	+125.0[2]	+173.0[2]
Methylbenzene	$C_6H_5CH_3$	l	178.1	383.7	0.867	+12.1	−3909.0	+110.6	+320.0
Ethylbenzene	$C_6H_5CH_2CH_3$	l	178.1	409.0	0.867	−13.1	−4563.9	+119.7	+255.2[2]
Amines									
Methylamine	CH_3NH_2	g	179.6	266.7	0.660[liq]	−23.0	−1072.0	+28.0	+243.3
Ethylamine	$CH_3CH_2NH_2$	g	182.0	289.7	0.683[liq]	−47.6	−1739.8	+37.0	+285.0
1−aminopropane	$CH_3CH_2CH_2NH_2$	l	190.0	320.9	0.719	−101.5	−2365.1	–	–
2−aminopropane	$CH_2CHNH_2CH_3$	l	177.9	305.5	0.688	−112.4	−2354.3	–	–
Phenylamine	$C_6H_6NH_2$	l	266.8	457.1	1.022	+31.3	−3397.0	–	–

		State (s,l,g)	M.Pt. K	B.Pt. K	Density g cm^{-3}	$\Delta H^\ominus_f$ kJ mol^{-1}	$\Delta H^\ominus_c$ kJ mol^{-1}	$\Delta G^\ominus_f$ kJ mol^{-1}	$S^\ominus$ $JK^{-1}mol^{-1}$
Halogenoalkanes									
Chloroethane	CH_3CH_2Cl	g	136.7	285.4	0.898[liq]	− 137.0	−1325.0	− 53.0	+ 276.0
Bromoethane	CH_3CH_2Br	l	154.5	311.4	1.460	− 90.5	−1427.4	–	–
Iodoethane	CH_3CH_2I	l	165.0	345.4	1.940	− 40.7	−1490.0	–	–
1,1,1−trichloroethane	CH_3CCl_3	l	242.7	347.2	1.339	− 177.2	− 1108.0	–	–
1−bromobutane	$CH_3CH_2CH_2CH_2Br$	l	160.0	374.7	1.279	− 143.8	−2716.5	–	–
Halogenoarenes									
Chlorobenzene	C_6H_5Cl	l	227.8	405.1	1.106	+ 11.1	− 3111.6	+ 93.6	–
Bromobenzene	C_6H_5Br	l	242.4	529.0	1.495	+ 60.5	–	+ 112.2	–
Iodobenzene	C_6H_5I	l	241.6	461.4	1.831	+ 114.5	−3193.0	+ 208.0	–
Alcohols									
Methanol	CH_3OH	l	179.2	337.5	0.791	−239.0	− 726.0	− 166.0	+ 127.0
Ethanol	CH_3CH_2OH	l	155.8	351.6	0.789	−277.7	− 1371.0	− 174.9	+ 160.7
Propan−1−ol	$CH_3CH_2CH_2OH$	l	146.6	370.2	0.803	−300.0	−2017.0	− 171.3	+ 196.6
Propan−2−ol	$CH_3CHOHCH_3$	l	183.5	355.4	0.786	− 317.9	−2005.8	− 180.3	+ 108.5
Butan−1−ol	$CH_3CH_2CH_2CH_2OH$	l	183.5	390.3	0.810	−327.4	−2673.0	− 169.0	+ 228.0
2−methylpropan−2−ol	$(CH_3)_3COH$	s/l	298.6	355.5	0.789	−359.0	−2643.8	–	–
Phenol	C_6H_5OH	s	316.1	454.9	1.073	− 165.0	−3053.4	− 50.9	+ 146.0
Cyclohexanol	$CH_2(CH_2)_4CHOH$	l	298.2	434.0	0.962	−349.0	−3727.0	− 134.2	–
Aldehydes									
Methanal	$HCHO$	g	181.0	252.0	0.815	− 108.7	− 570.6	− 110.0	+ 219.0
Ethanal	CH_3CHO	g	152.1	293.9	0.783	− 166.4	− 1167.0	− 134.0	+ 160.2[2]
Propanal	CH_3CH_2CHO	l	192.0	321.9	0.797	− 217.0	−1820.8	− 142.1	–
Butanal	$CH_3CH_2CH_2CHO$	l	174.1	348.7	0.817	− 241.2	−2476.0	−306.4	–
Benzaldehyde	C_6H_5CHO	l	247.1	451.1	1.050	− 86.8	−3520.0	–	–
Ketones									
Propanone	CH_3COCH_3	l	177.7	329.2	0.790	− 216.7	− 1821.0	− 154.8	+ 295.0[gas]
Butanone	$CH_3CH_2COCH_3$	l	186.8	352.7	0.805	−276.0	−2441.5	− 156.0	–
Phenylethanone	$C_6H_5COCH_3$	l	293.6	475.7	1.028	− 142.5	−4138.0	–	–
Carboxylic acids									
Methanoic acid	$HCOOH$	l	281.5	374.0	1.220	−425.0	− 262.8	−346.0	+ 129.0
Ethanoic acid	CH_3COOH	l	289.7	391.0	1.049	−488.3	− 876.0	−392.0	+ 160.0
Chloroethanoic acid	$ClCH_2COOH$	s	336.1	462.0	1.404	–	− 716.0	–	–
Dichloroethanoic acid	$Cl_2CHCOOH$	l	286.6	467.1	1.566	–	–	–	–
Trichloroethanoic acid	Cl_3CCOOH	s	331.1	470.6	1.617	− 514.0	− 388.0	–	–
1−aminoethanoic acid	NH_2CH_2COOH	s	535[dec]	dec	1.607	−528.6	− 981.0	–	–
2−hydroxypropanoic acid	$CH_3CHOHCOOH$	l	289.8	dec	1.206	−694.0	−1344.0	–	–
Benzenecarboxylic acid (Benzoic acid)	C_6H_5COOH	s	395.0	522.0	1.266[288K]	−390.0	−3227.0	−245.0	+ 167.0
Carboxylic acids and derivatives									
Ethanoyl chloride	CH_3COCl	l	161.0	324.0	1.105	−273.0	–	−208.0	+ 200.8
Ethanamide	CH_3CONH_2	s	355.3	494.3	1.159	− 317.0	− 1182.0	–	–
Ethanoic anhydride	$(CH_3CO)_2O$	l	199.9	413.0	1.082	−637.2	−1794.2	–	–
Methylethanoate	CH_3COOCH_3	l	175.0	330.2	0.933	−386.0	− 972.5	–	–
Ethylethanoate	$CH_3COOCH_2CH_3$	l	189.4	350.1	0.972	−479.3	−2238.0	–	–

		State (s,l,g)	M.Pt. K	B.Pt. K	Density g cm^{-3}	$\Delta H^{\ominus}_{f}$ kJ mol^{-1}	$\Delta H^{\ominus}_{c}$ kJ mol^{-1}	$\Delta G^{\ominus}_{f}$ kJ mol^{-1}	$S^{\ominus}$ $JK^{-1}mol^{-1}$
Nitriles									
Propanonitrile	CH_3CH_2CN	l	181.1	370.3	0.772	+147.2	−1910.5	–	–
Miscellaneous									
Benzenesulphonic acid	$C_6H_5SO_3H$	s	323.0	–	–	–	–	–	–
Cholestrol	$C_{27}H_{44}OH$	s	421.6	633.1^{dec}	1.067	–	–	–	–
Methyl−2−nitrobenzene	$CH_3C_6H_4NO_2$	l	270.1	493.0	1.163	–	–	–	–
Nitrobenzene	$C_6H_5NO_2$	l	278.8	483.9	1.203–	+12.4	−3094.0	+141.6	–
1,3−dinitrobenzene	$C_6H_4(NO_2)_2$	s	363.0	564.0	1.580	–	–	–	–
Sucrose	$C_{12}H_{22}O_{11}$	s	458.0	–		−2226.0	−5644.0	–	–

c) pK_{in} and pH range values for common indicators

Indicator	Colour in Acid	Colour in Alkali	pK_{in}	pH Range
Thymol blue (acid)	Red	Yellow	1.7	1.2 – 2.8
Methyl orange	Red	Yellow	3.7	3.1 – 4.4
Bromphenol blue	Yellow	Blue	4.0	3.0 – 4.6
Methyl red	Red	Yellow	5.1	4.2 – 6.3
Bromothymol blue	Yellow	Blue	7.0	6.0 – 7.6
Phenol red	Yellow	Red	7.9	6.8 – 8.4
Thymol blue (base)	Yellow	Blue	8.9	8.0 – 9.6
Phenolphthalein	Colourless	Pink	9.3	8.2 – 10.0

d) Equilibrium constants (pK_a) for some substances in aqueous solution.

	Formula	pKa
<u>Organic Acids</u>		
Methanoic acid	$HCOOH$	3.73
Ethanoic acid	CH_3COOH	4.76
Propanoic acid	CH_3CH_2COOH	4.87
Butanoic acid	$CH_3CH_2CH_2COOH$	4.82
Chloroethanoic acid	$ClCH_2COOH$	2.86
Dichloroethanoic acid	$Cl_2CHCOOH$	1.29
Trichloroethanoic acid	Cl_3CCOOH	0.65
Bromoethanoic acid	$BrCH_2COOH$	2.90
1–aminoethanoic acid (glycine)	NH_2CH_2COOH	9.87
Benzenecarboxylic acid (benzoic acid)	C_6H_5COOH	4.20
2–chlorobenzenecarboxylic acid	ClC_6H_4COOH	2.94
3–chlorobenzenecarboxylic acid	ClC_6H_4COOH	3.83
4–chlorobenzenecarboxylic acid	ClC_6H_4COOH	3.99
<u>Alcohols</u>		
Methanol	CH_3OH	15.50
Ethanol	CH_3CH_2OH	16.2
Phenol	C_6H_5OH	10.00
2,4,6–trichlorophenol (TCP)	$Cl_3C_6H_2OH$	7.60
<u>Amines</u>		
Ammonia	NH_3	9.25
Methylamine	CH_3NH_2	10.64
Dimethylamine	$(CH_3)_2NH$	10.72
Trimethylamine	$(CH_3)_3N$	9.80
Phenylamine	$C_6H_5NH_2$	4.62
<u>Inorganic acids</u>		
Carbonic acid	$(H_2O + CO_2)$	6.35
Chloric(I) acid	$HClO$	7.43
Hydrofluoric acid	HF	3.25
Nitric acid	HNO_3	−1.60
Nitrous acid	HNO_2	3.33
Sulphurous acid	H_2SO_3	1.82
<u>Miscellaneous</u>		
Aluminium ion	$Al(H_2O)_6^{3+}$	5.00
Ammonium ion	NH_4^+	9.25
Hydrogencarbonate ion	HCO_3^-	10.32
Hydrogensulphate ion	HSO_4^-	2.00
Water	H_2O	14.00

e) Correlation chart of infrared absorption bands

This chart shows characteristic ranges of infrared absorptions due to stretching and bending motions in organic molecules.

Stretching vibration

Bending vibration

$$\text{Wavenumber (cm}^{-1}\text{)} = \frac{1}{\text{wavelength (cm)}}$$

Stretching

Bond	Compound type	Wavenumber / cm^{-1}
C–Cl		800–600
C–Br		600–500
C–I		500
C–O	alcohols, ethers, carboxylic acids, esters	1310–1000
C–F		1400–1000
C=O	aldehydes, ketones, acids & derivatives	1850–1630
C=C	alkene	1680–1620
C=C	arene	1600–1450
C≡N	nitrile	2260–2210
C≡C	alkynes	2260–2100
C–H	aldehyde	2900–2700
C–H	alkane	2960–2850
C–H	alkene	3095–3010
C–H	arenes	3030
C–H	alkyne	3300
N–H	amine	3560–3300
N–H	amide	3500–3140
* O–H	alcohol/phenol	3670–3580
+ O–H	alcohols/phenols	3550–3200
+ O–H	carbonyl compounds	3500–2500

Bending

Bond	Compound type	Wavenumber / cm^{-1}
C–H	alkyne	630
C–H	arene	880–670
C–H	alkene	1420–750
O–H	alcohols/phenols	1200–1050
C–H	alkane	1485–1340

Wavenumber / cm^{-1}: 4000, 3000, 2000, 1000, 400

* no hydrogen bonding
\+ hydrogen bonding

f) Sample infrared spectra

Infrared spectrum of ethanol (liquid film)

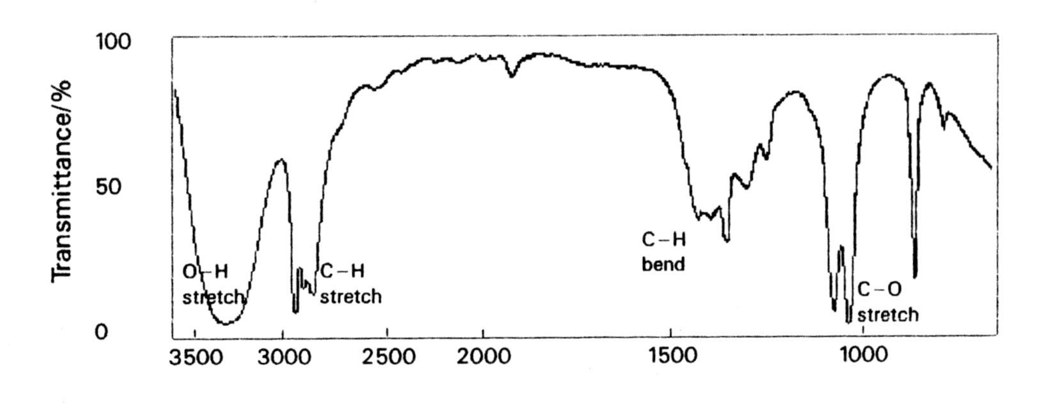

Infrared spectrum of methylbenzene (liquid film)

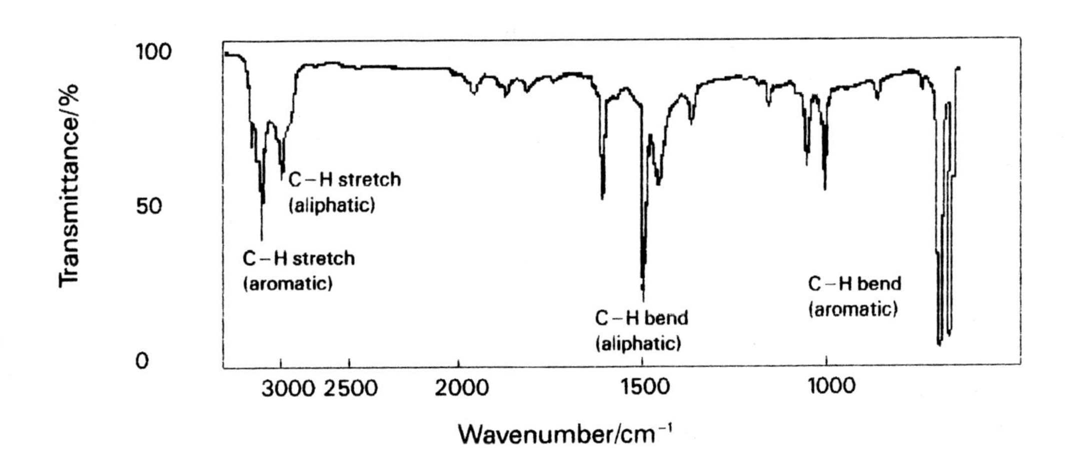

PERIODIC TABLE

Transition metals

	1 (I)	2 (II)											3 (III)	4 (IV)	5 (V)	6 (VI)	7 (VII)	0
1					$^{1}_{1}$H Hydrogen													$^{4}_{2}$He Helium
2	$^{7}_{3}$Li Lithium	$^{9}_{4}$Be Beryllium											$^{11}_{5}$B Boron	$^{12}_{6}$C Carbon	$^{14}_{7}$N Nitrogen	$^{16}_{8}$O Oxygen	$^{19}_{9}$F Fluorine	$^{20}_{10}$Ne Neon
3	$^{23}_{11}$Na Sodium	$^{24}_{12}$Mg Magnesium											$^{27}_{13}$Al Aluminium	$^{28}_{14}$Si Silicon	$^{31}_{15}$P Phosphorus	$^{32}_{16}$S Sulphur	$^{35.5}_{17}$Cl Chlorine	$^{40}_{18}$Ar Argon
4	$^{39}_{19}$K Potassium	$^{40}_{20}$Ca Calcium	$^{45}_{21}$Sc Scandium	$^{48}_{22}$Ti Titanium	$^{51}_{23}$V Vanadium	$^{52}_{24}$Cr Chromium	$^{55}_{25}$Mn Manganese	$^{56}_{26}$Fe Iron	$^{59}_{27}$Co Cobalt	$^{59}_{28}$Ni Nickel	$^{63.5}_{29}$Cu Copper	$^{65}_{30}$Zn Zinc	$^{70}_{31}$Ga Gallium	$^{73}_{32}$Ge Germanium	$^{75}_{33}$As Arsenic	$^{79}_{34}$Se Selenium	$^{80}_{35}$Br Bromine	$^{84}_{36}$Kr Krypton
5	$^{85}_{37}$Rb Rubidium	$^{88}_{38}$Sr Strontium	$^{89}_{39}$Y Yttrium	$^{91}_{40}$Zr Zirconium	$^{93}_{41}$Nb Niobium	$^{96}_{42}$Mo Molybdenum	$^{99}_{43}$Tc Technetium	$^{101}_{44}$Ru Ruthenium	$^{103}_{45}$Rh Rhodium	$^{106}_{46}$Pd Palladium	$^{108}_{47}$Ag Silver	$^{112}_{48}$Cd Cadmium	$^{115}_{49}$In Indium	$^{119}_{50}$Sn Tin	$^{122}_{51}$Sb Antimony	$^{128}_{52}$Te Tellurium	$^{127}_{53}$I Iodine	$^{131}_{54}$Xe Xenon
6	$^{133}_{55}$Cs Caesium	$^{137}_{56}$Ba Barium		$^{178.5}_{72}$Hf Hafnium	$^{181}_{73}$Ta Tantalum	$^{184}_{74}$W Tungsten	$^{186}_{75}$Re Rhenium	$^{190}_{76}$Os Osmium	$^{192}_{77}$Ir Iridium	$^{195}_{78}$Pt Platinum	$^{197}_{79}$Au Gold	$^{201}_{80}$Hg Mercury	$^{204}_{81}$Tl Thallium	$^{207}_{82}$Pb Lead	$^{209}_{83}$Bi Bismuth	$^{209}_{84}$Po Polonium	$^{210}_{85}$At Astatine	$^{222}_{86}$Rn Radon
7	$^{223}_{87}$Fr Francium	$^{226}_{88}$Ra Radium		$^{261}_{104}$Unq Unnil-quadium	$^{262}_{105}$Unp Unnil-pentium	$^{263}_{106}$Unh Unnil-hexium	$^{262}_{107}$Uns Unnil-septium	$_{108}$Uno Unnil-octium	$_{109}$Une Unnil-ennium									

$^{139}_{57}$La Lanthanum	$^{140}_{58}$Ce Cerium	$^{141}_{59}$Pr Praseodymium	$^{144}_{60}$Nd Neodymium	$^{147}_{61}$Pm Promethium	$^{150}_{62}$Sm Samarium	$^{152}_{63}$Eu Europium	$^{157}_{64}$Gd Gadolinium	$^{159}_{65}$Tb Terbium	$^{162}_{66}$Dy Dysprosium	$^{165}_{67}$Ho Holmium	$^{167}_{68}$Er Erbium	$^{169}_{69}$Tm Thulium	$^{173}_{70}$Yb Ytterbium	$^{175}_{71}$Lu Lutetium
$^{227}_{89}$Ac Actinium	$^{232}_{90}$Th Thorium	$^{231}_{91}$Pa Protactinium	$^{238}_{92}$U Uranium	$^{237}_{93}$Np Neptunium	$^{244}_{94}$Pu Plutonium	$^{243}_{95}$Am Americium	$^{247}_{96}$Cm Curium	$^{247}_{97}$Bk Berkelium	$^{251}_{98}$Cf Californium	$^{252}_{99}$Es Einsteinium	$^{257}_{100}$Fm Fermium	$^{258}_{101}$Md Mendelevium	$^{259}_{102}$No Nobelium	$^{260}_{103}$Lw Lawrencium

Association for Science Education, *Chemical Nomenclature, Symbols and Terminology*, 1984.
Association for Science Education, *SI Units, Signs, Symbols and Abbreviations*, 1981.
Baughan E.C., *Trans. Faraday Soc.*, 1961, 57, 1863.
Bushell J. and Nicholson P., *Biology Alive*, Collins Educational, 1985.
Chang R., *Physical Chemistry with Applications to Biological Systems*, 2nd Edition, Macmillan, 1981.
Colthup N.B., Daly L.H. and Wiberley S.E., *Introduction to Infra-red and Raman Spectroscopy*, Academic Press, 1964.
Cotton F.A. and Wilkinson G., *Advanced Inorganic Chemistry*, 4th Edition, John Wiley and Sons, 1980.
Cross A.D., *Introduction to Practical Infra-red Spectroscopy*, Butterworth and Co. (Publishers) Ltd., 1964.
Dodd R.E., *Chemical Spectroscopy*, Elsevier Publishing Company, 1962.
Durrant P.J., *General and Inorganic Chemistry*, Longmans, 1964.
Emsley J., *The Elements*, Clarendon Press, Oxford, 1989.
Fox B.A. and Cameron A.G., *Food Science - A Chemical Approach*, Unibooks, 1980.
Ginsberg A.P. and Miller J.M., *J. Inorg. Nucl. Chem.*, 1958, 7 351.
Green N.P.O., Stout G.W. and Taylor D.J., *Biological Science, Book One*, Cambridge University Press, 1985.
Greenwood N. and Earnshaw A., *Chemistry of the Elements*, Pergamon, 1984.
The Guinness Book of Answers, Guiness Superlatives Ltd., 1976.
Hanssen M., *E for Additives*, Thorsons Publishers Ltd., 1985.
Heaton C.A., *An Introduction to Industrial Chemistry*, Leonard Hill, 1984.
Hotop H. and Lineberger W.C., *J. Phys. Chem. Ref. Data*, 1975, 4, 539.
ILPAC Unit 13, John Murray Ltd., 1984.
Kaye G.W.C. and Laby T.H., *Tables of Physical and Chemical Constants*, 14th edition, Longman, 1973.
Liptrot G.F., *Modern Inorganic Chemistry*, Mills and Boon, 1980.
Lowry A.G. and Siekevitz P., *Cell Structure and Function*, 2nd Edition, Holt International, 1970.
McElory W.D., *Cellular Physiology and Biochemistry*, Prentice Hall Inc., 1961.
Murray P.R.S., *Principles of Organic Chemistry*, 2nd Edition, Heinemann Educational Books, 1980.
Norman R.O.C. and Waddington D.J., *Modern Organic Chemistry*, 4th Edition, Bell and Hyman, 1983.
Nuffield Advanced Science Book of Data, Longman, 1984.
Nuffield Advanced Science, Chemistry (Revised), Students/Teachers Books 1 and 2, 1985.
Parkes G.D., *Mellor's Modern Inorganic Chemistry*, Longman, 1960.
Perel'man F.M., *Rubidium and Cesium*, Macmillan, 1965.
Pine S.H., Hendrickson J.B., Cram D.J. and Hammond G.S., *Organic Chemistry*, 4th Edition, McGraw-Hill, 1984.
Ramsden E.N., *A-level Chemistry*, 2nd Edition, Stanley Thornes Ltd., 1990.
Roberts M.B.V., *Biology - A Functional Approach*, 4th Edition, Nelson, 1986.
Science Data Book, The Open University, Oliver and Boyd, 1984.
Sharp D.W.A., *The Penguin Dictionary of Chemistry*, Penguin Books, 1983.
Stark J.G. and Wallace H.G., *Chemistry Data Book*, 2nd Edition, John Murray, 1982.
Uvarov E.B., Chapman D.R. and Isaacs A., *The Penguin Dictionary of Science*, 5th Edition, Penguin Books, 1979.
Warn J.R.W., *Concise Chemical Thermodynamics*, Van Nostrand Reinhold Company, 1969.
Wilson J.G. and Newall A.B., *General and Inorganic Chemistry*, 2nd Edition, Cambridge University Press, 1970.
Wood E.J. and Pickering W.R., *Introducing Biochemistry*, John Murray, 1982.